SpringerBriefs in Applied Sciences and Technology

For further volumes:
http://www.springer.com/series/8884

K. S. K. Weranga · Sisil Kumarawadu
D. P. Chandima

Smart Metering Design and Applications

Springer

K. S. K. Weranga
Sisil Kumarawadu
D. P. Chandima
Department of Electrical Engineering
University of Moratuwa
Moratuwa
Sri Lanka

ISSN 2191-530X ISSN 2191-5318 (electronic)
ISBN 978-981-4451-81-9 ISBN 978-981-4451-82-6 (eBook)
DOI 10.1007/978-981-4451-82-6
Springer Singapore Heidelberg New York Dordrecht London

Library of Congress Control Number: 2013942121

Printed on acid-free paper

Springer is part of Springer Science+Business Media (www.springer.com)

Preface

Taking into account the present day trends and the requirements, this brief focuses on smart metering of electricity for next generation energy efficiency and conservation. The contents include discussions on smart grid and smart metering concepts, existing technologies and systems as well as design and implementation of smart metering schemes together with detailed examples.

In the first chapter the definitions and the concept of smart grid are discussed. The drawbacks of existing grid systems and the importance of developing the smart grids are also explained. The structure of the smart metering systems and its benefits, relevance of the demand management integrated to smart metering, worldwide smart metering projects, and the trends of smart metering in some countries are discussed.

Chapter 2 covers the evolution of electricity meters from typical electromechanical meters to modern smart meters. The theory, operation principles, and drawbacks of typical electromechanical meters are addressed. The operating mechanism of reactive energy meters and maximum demand meters are also addressed. The development of solid state meters and conversion of the technology to smart meters are pointed out. This chapter reveals the advantages of smart meters in modern energy measurement and the basic hardware structure of a modern smart meter. A broader discussion is done about the hardware components inside a smart meter including voltage and current sensors, power supplies, energy measurement, microcontroller, real time clock, and communication protocols. The standards used for smart meters are also highlighted.

Chapter 3 consists of functions, calculations, and operations inside standard energy measurement chips commonly available in the market. Signal conditioning, analog to digital conversion, and electrical parameter calculations are described step by step where the reader can get an overall idea on digital energy measurement.

Design and implementation examples are covered in Chap. 4 including the selection of the component parts from a range of off-the-shelf options available, displaying per phase and three phase quantities on LCD at the meter side, automatic meter reading, power quality and exported energy measuring capability, as well as consumer notifications. The methods used for consumer notifications highlight the consumption details including the electricity cost, average energy usage, average daily cost, predicted electricity cost for the month, and the number of remaining days for specified credit levels.

Chapter 5 focuses on the short-term electricity demand forecasting and how it contributes to demand side load management. Several types of short-term demand forecasting methods are summarized. A case study is done by selecting an industrial consumer to show how the short-term demand forecasting can be applied with smart metering. Polynomial regression is used to forecast the demand by taking the apparent power sample points from smart meters. This methodology can be used to forecast the demand, calculate the maximum demand, and to control the demand side loads. With the aid of smart meters, warning signals can be generated and sent to the consumer to reduce the demand and save the electricity bill.

Chapter 6 presents the applications of smart meters in today's world. The integration of smart metering in distributed generation, voltage, and reactive power monitoring in distributed grid, enhancement of HVAC system performance, and demand side load management are discussed. The residential load controlling through smart metering, the integration of smart appliance controllers, demand side primary frequency control, optimal energy management under time varying tariff structures, and their impacts on energy conservation and electricity cost reduction are summarized with modern world examples.

The authors would like to express their gratitude to Analog Devices, Inc., Allegro MicroSystems, Inc., STMicroelectronics, Taehwatrans Co. Ltd for allocating the permission to include some content and figures which were originally presented in their data sheets and websites.

Contents

1 Smart Grid and Smart Metering 1
1.1 Introduction 1
1.2 Smart Grid 2
1.3 Smart Metering 6
1.4 Energy Saving Through Smart Metering 8
1.5 Techniques Used for Demand Side Management 8
1.5.1 Educating the Consumer 8
1.5.2 Incenting the Consumer 9
1.5.3 Controlling the Consumer 9
1.6 Carbon Benefits Through Smart Metering 10
1.7 Smart Metering Projects 10
1.7.1 Smart Metering Projects in Italy 10
1.7.2 Smart Metering Projects in USA 11
1.7.3 Smart Metering Projects in Asia Pacific 12
1.7.4 Smart Metering Projects in Canada 12
1.7.5 Smart Metering Projects in UK 12
1.7.6 Smart Metering Projects in China 13
1.7.7 Smart Metering Projects in Japan 13
References 14

2 Evolution of Electricity Meters 17
2.1 Introduction 17
2.2 Operation Principles of an Electromechanical Energy Meter . . . 18
2.3 Drawbacks of the Electromechanical Energy Meters 22
2.4 Reactive Energy Meters 23
2.5 Maximum Demand Meters 24
2.6 Electronic Meters 25
2.7 Smart Meters 26
2.8 The Hardware Structure of a Smart Meter 27
2.8.1 The Voltage Sensing Unit 28
2.8.2 The Current Sensing Unit 29
2.8.3 Power Supply 33
2.8.4 Energy Measurement Unit 34

2.8.5 Microcontroller 35
2.8.6 Real Time Clock 35
2.8.7 Communicating Systems 36
2.9 Smart Meter Standards 36
References 37

3 Basic Functionalities Inside an Energy Measurement Chip 39
3.1 Introduction 39
3.2 Signal Conditioning 40
3.3 Analog to Digital Conversion 41
3.4 Basic Electrical Parameters Calculation in a Sinusoidal Single Phase System 44
3.4.1 RMS Calculation 44
3.4.2 Active Power and Energy Calculation 47
3.4.3 Reactive Power and Energy Calculation 50
3.4.4 Apparent Power and Energy Calculation 53
3.5 Power Quality Measurements 53
3.6 Electrical Parameters Calculation in a Non-sinusoidal Single Phase System 54
3.6.1 Harmonics and Energy Measurement in ADE 7880 Energy Measurement Chip 59
3.7 Four Quadrant Operation 62
3.8 Frequency Outputs 63
3.9 Communication 64
References 64

4 Smart Meter Prototype Design 65
4.1 Introduction 65
4.2 Basic Operation of ADE7758 Energy Chip 66
4.3 Current Sensing Unit 67
4.4 Voltage Sensing Unit 68
4.5 Calculations 70
4.5.1 RMS Voltage and RMS Current Calculation 70
4.5.2 Active Power and Energy Calculation 75
4.5.3 Reactive Power and Energy Calculation 77
4.5.4 Apparent Power Calculation 77
4.5.5 Power Factor Calculation 79
4.5.6 Frequency Calculation 79
4.5.7 Power Quality Measurements 80
4.6 Energy Chip and MCU Communication 80
4.7 Power Supply Unit 80
4.8 Data Backup 81
4.9 Real Time Clock 81

4.10 Smart Meter Firmware Development 82
4.10.1 The Main Program 82
4.10.2 TIMER0 Interrupt Routine 83
4.10.3 INTCON Interrupts Routine 86
4.11 Data Transmission 86
4.12 The Serial Port Listener 88
4.13 The Graphical User Interface 89
4.14 Bill Generation, Cost Prediction and Customer Update 89
4.15 Electricity Consumption Details 91
References 93

5 Short-Term Electricity Demand Forecasting and Warning Signal Generation 95
5.1 Introduction 95
5.2 Electricity Demand Forecasting Methods 96
5.3 A Case Study on How the STDF is Applied to Reduce the Maximum Demand Charge 99
References 113

6 Smart Metering Applications 115
6.1 Introduction 115
6.2 Distributed Generation 116
6.3 Voltage Control in Smart Grid 117
6.4 Load Control 118
6.5 Demand Side Primary Frequency Variation with Load Control 120
6.6 Enhance the HVAC System Performance 120
6.7 Worldwide Load Control Programs 121
6.8 Optimal Energy Management 121
References 123

Appendix 1: Schematic of Energy Measurement Chip 125

Appendix 2: Schematic of PIC 18F452/ RTC/ LCD 127

Appendix 3: Schematic of SIM900 GSM Module 129

Appendix 4: Schematic of Power Supply 131

Appendix 5: Some Useful Functions for Software Development 133

Appendix 6: Mathlab Code for Simulation of Fig. 5.5 139

Abbreviations

A	Ampere
ADC	Analog to Digital Conversion
AEIC	Association of Edison Illuminating Companies
AMI	Advanced Metering Infrastructure
AMR	Automatic Meter Reading
ANNs	Artificial Neural Networks
ANSI	American National Standards Institute
AR	Autoregressive
ARMA	Autoregressive Moving Average
AT	Asynchronous Terminal
CT	Current Transformer
DAC	Digital to Analog Conversion
DDC	Dynamic Demand Control
DG	Distributed Generation
DSP	Digital Signal Processor
FCC	Federal Communication Commission
FCDM	Frequency Controlled Demand Management
GDP	Gross Domestic Product
GPRS	General Packet Radio Services
GSM	Global System for Mobile Communications
GUI	Graphical User Interface
HAN	Home Area Network
HVAC	Heating, Ventilation, and Air-Conditioning
IC	Integrated Circuit
IEC	International Electro-technical Commission
IEEE	Institute of Electrical and Electronic Engineers
IHD	In-Home Display
ISO	Independent System Operator
I2C	Inter-Integrated Circuit
kVA	kilo Volt Ampere
kVAh	kilo Volt Ampere hour
kVAmin	kilo Volt Ampere minutes
kvarh	kilo var hour
kW	kilo Watt

kWh	kilo Watt hour
LCD	Liquid Crystal Display
LKR	Sri Lanka Rupees
LPF	Low-Pass Filter
MA	Moving Average
MCU	Micro Controller Unit
MOV	Metal Oxide Varistor
MWh	Mcga Watt hour
NAN	Neighborhood Area Network
NEMA	National Electrical Manufacturer Association
NIST	National Institute of Standards and Technology
NNs	Neural Networks
OLTC	On-Load Tap Changer
PG&E	Pacific Gas and Electric
PHEV	Plug-in Hybrid Electric Vehicle
PLC	Power Line Communication
PV	Photovoltaic
RF	Radio Frequency
RMS	Root Mean Square
RTC	Real Time Clock
RTO	Regional Transmission Organization
RTP	Real Time Pricing
SCADA	Supervisory Control and Data Acquisition
SGIP	Smart Grid Interoperability Panel
SMS	Short Message Service
SPI	Serial Peripheral Interface
STDF	Short-Term Demand Forecasting
STEDF	Short-Term Electricity Demand Forecasting
THD	Total Harmonic Distortion
TOU	Time Of Use
V	Volt
W	Watt
WAN	Wide Area Network

Chapter 1
Smart Grid and Smart Metering

Abstract The concept, structure and benefits of the smart gird and the smart metering are discussed in this chapter. The old gird system, its drawbacks, and the importance of a smart grid in today's world are remarked. The structure of smart metering systems and its benefits including energy efficiency and conservation are highlighted. Relevance of demand management integrated with smart metering is reported. Worldwide smart metering projects, their outcomes, and trends of smart metering in some countries are also discussed.

1.1 Introduction

Electricity or the electric power has become the dominant variable of any economy. Electric energy is used to power up the domestic, commercial and industrial loads as a key source of energy in today's world. The grid system that has been developed over the last 70 years mainly established upon three pillars that are generation, transmission and distribution. Typically large centralized generator systems are used to convert the other source of energy such as thermal, hydro, and coal into electric energy. This electric energy is then transmitted via high voltage transmission lines over long distances. The power coming from this high voltage lines is stepped down at substations and passed through low voltage transformers to power up the end user loads. The reliability of the system is guaranteed by excess capacity in the system with unidirectional flow of electricity. A communication method is used between the generator units and the transmission grid even though it has limited functionality. Moreover the distribution network also has limited controllability since it hasn't enough communication infrastructures.

Electricity consumers are asking for better customer service, high accuracy in energy measurement, and healthy power supply along with timely data delivery. Limited information about the electricity usage (usual monthly electricity bill)

K. S. K. Weranga et al., *Smart Metering Design and Applications*,
SpringerBriefs in Applied Sciences and Technology,
DOI: 10.1007/978-981-4451-82-6_1, © The Author(s) 2014

provided to the end users make fewer incentives to adapt their energy usage to save the electricity bill. On the other hand the electromechanical meters are incapable of displaying the real time information of energy usage and time varying prices at the consumer side.

The ever increasing demand for electricity has caused several problems to power utilities and governments in many countries. The power system expanded rapidly from 1950s mainly in the United States of America and some European countries. A rapid growth of distributed generation (DG) has also been observed. Each day, more renewable energy sources are added to the system apart from the centrally generated ones. Since wind power and solar power are highly variable, more sophisticated control systems are needed to facilitate the grid [1].

Moreover the old grid system is not readily configurable to rapidly changing demand patterns. It suffers from many shortcomings like poor efficiency, lack of reliability, lack of energy buffering, high cost of energy consumption, low fault detection speed, carbon pollution and insufficient interaction between the consumer and the grid company. Transmission and distribution equipments are out of date and should be replaced. Meanwhile the government and regulations are forcing the utilities for more competition, efficiency, low price for electricity, and green energy. Therefore the necessity for an advanced grid system is identified.

Research and developments in power system engineering have contributed in developing of a reliable and highly efficient grid system which supports DG, security, reliability and two-way interaction. The improvements in the electronic communication technology are also used to resolve the limitations of the old grid system [2]. The incorporation of modern telecommunication technologies has established a reliable communication link all over the grid system making it easy to monitor and control. This communication infrastructure is used to monitor and control the power usage at different locations in the grid system. An advanced metering infrastructure (AMI) is also needed to view and analyse the demand patterns on a per-user basis. The replacement of electromechanical meters with smart meters along with domestic load controllers is identified for better energy conservation at the consumer side.

Ultimately the concept of smart grid is raised concerning the drawbacks of the old grid system and the necessity for a new intelligent grid system that has improved reliability, security, and efficiency.

1.2 Smart Grid

Even though there is no exact definition for "Smart Grid", we can say that smart grid is basically an intelligent electricity delivery system combined with modern digital and information technology, which provides efficiency, security, reliability and more benefits for both utilities and consumers.

There are several other definitions of Smart grid.

According to the Smart Grid Communications Task Force [3]:

Smart Grid is a term used for an advanced electricity delivery system that is integrated with modern digital and information technology to provide improved reliability, security, efficiency, and ultimately lower cost of the utility services to the user.

The US Department of Energy [4] defines:

A smart grid uses digital technology to improve reliability, security, and efficiency (both economic and energy) of the electric system from large generation, through the delivery systems to electricity consumers and a growing number of distributed-generation and storage resources.

The Department of Environment, Water, Heritage and the Art of Australian Government declares [5]:

Smart grids combine advanced communication and metering infrastructure with existing energy networks to enable a combination of grid-side and customer applications to deliver a more efficient and robust network

The vision of smart grid is to modernize the existing grid system for better efficiency, reliability and security. It will be expensive to develop and deploy a smart grid. However in the long run, a smart grid will be a good option for sustainability. According to the Ref. [6], the benefits associated with the smart grid include:

- More efficient transmission of electricity
- Quicker restoration of electricity after power disturbances
- Low operation and maintenance cost for the utilities
- Lower power cost for electricity consumers
- Increased integration of large-scale renewable energy systems
- Improved security

Many smart grid technologies are being adapted for use in grid operations. In general, smart grid technology can be grouped into five categories- integrated communications, sensing and measurement, smart metering, phasor measurement units and advanced components [12]. These technologies will work with the electrical grid to respond digitally for quickly changing electricity demand. Figure 1.1 shows the basic services integrated with a smart grid system [7].

The sensing and measurement system can be divided into three sections that are AMI, phasor measurement and distributed weather sensing. AMI ensures the two way interaction between the consumer and the utility. It provides real time electricity pricing, electricity usage, electricity cost, outage detection and accurate load characterization. AMI is basically integrated with smart meters, in-home displays, and load controllers. Phasor measurement units are used to monitor the waveforms of the system, to measure the health of the system, to increase the reliability, and to prevent the power outages. Distributed weather sensing units provide solar irradiance, wind speed and temperature measurements to forecast and manage the renewable energy [8].

Integrated communications and security system allow the user and the utility to manage the various intelligent electronic devices such as load controllers, smart

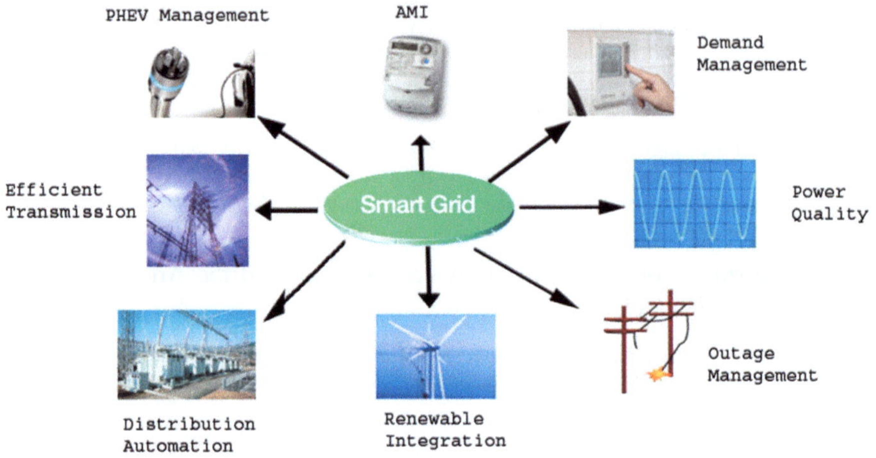

Fig. 1.1 Services integrated with a smart grid system

meters and sensors in a secure and reliable way. This system is capable of handling near or real time information to improve the reliability, security, efficiency of power delivery and usage. The commonly used communication links are power line communication, broad band cables, Wi-fi, ZigBee, GPRS, 3G, and radio frequencies.

Moreover a good demand response can be achieved through a smart grid. The consumer can adjust his load by responding to various price signals appearing on the smart meter. He can also set smart appliances to respond to these price signals to minimize the cost of electricity. This kind of active demand management can help save electricity bill. Besides, it will increase the system load factor. Smart grid promotes localized power generation due to its capability of measuring power in both directions. This allows small localized generators to push their unused locally generated power back to the system. They are paid accurately for the power generated at the same time. Nevertheless customers who use electric vehicles can charge batteries at off-peak hours based on the expected prices and car use patterns. Not only they can charge the vehicles, but also they can sell back or return stored energy during on-peak hours with the bidirectional metering [9].

Figure 1.2 shows a conceptual model of smart grid which has seven main functional areas. The customers or the end users can be residential, commercial or industrial electricity consumers. They have the capability of generate, store and manage the electricity. Consumers are updated with information flaws from markets, operators and service providers. Smart meters, appliance controllers, HVAC (heat ventilation air-conditioning), In-home display units, distributed energy resources, production process devices and electric vehicles are the main physical equipments used at the consumer side.

Markets are the operators and participants in electricity markets. They are hardware providers, software firms, telecommunication companies, and network

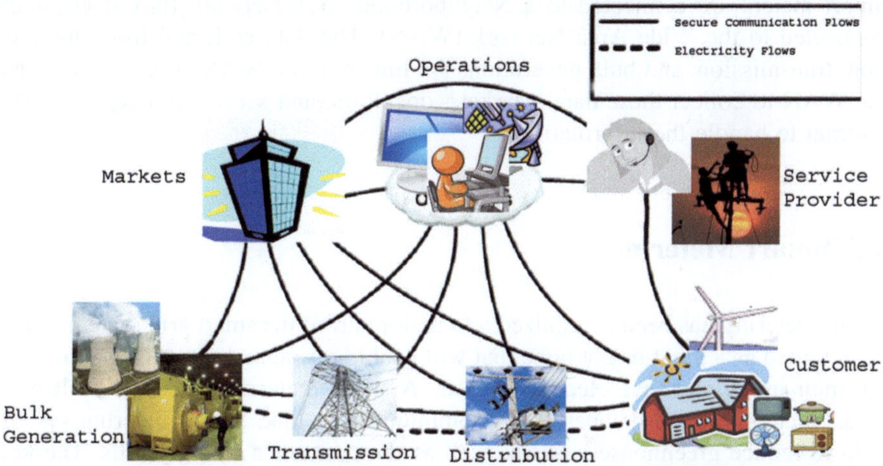

Fig. 1.2 A conceptual model of Smart Grid

companies. Service providers are the organizations who handle all third-party operations among the domains. The main services are billing, customer management, installation and maintenance, home management, and emergency services.

Operations do the management and control of movement of electricity in the smart grid. Basic operational functions include monitoring, controlling, reporting, and supervising. Here all the substations, customer premises networks and intelligent field devices are connected using a two-way communications network. The main physical equipments associated to operations are ISO/RTO SCADA, Distribution SCADA, Transmission SCADA, and Management Stations.

The bulk generation includes the generation of electricity in bulk quantity using renewable and non-renewable sources. It may also store energy for later distribution. The transmission domain is to carry the electricity over long distances. Bulk generation and transmission domain equipped with plant controls, distributed energy resources, transmission and substation controllers.

The distribution domain performs three main tasks which are distribution of electricity to customers, connection of smart meters and all intelligent field devices, and management of energy storage facilities [10].

The information flows between these seven areas and the electric power flaws through the transmission and the distribution network. Even the consumer can generate his own electricity and sell back excess power to the grid.

The communication infrastructure is a key element to build a high-performance smart grid. Each domain has several communication links to several other domains to transfer the data. On the other hand there are network infrastructures inside each domain to serve need of the domain.

At the customer premises the Home Area Network (HAN) is used to communicate with appliances, smart meters, electric vehicles and local generators.

Smart meters are connected to a Neighborhood Area Network (NAN) and then connected to the Wide Area Network (WAN). The data collected from distribution, transmission, and bulk generations are linked to the WAN. The operators use the WAN to collect these data. Markets, operators, and service providers use the internet to handle the information.

1.3 Smart Metering

Smart metering has been recognized as a major part of the smart grid system. It has been touted as a great bright hope that will enable residential electric customers to cut their usage and save electricity costs. A smart metering system is built with smart meters, control devices and a communication link. Smart metering system help to reduce greenhouse gases as well as the consumer monthly bills. The key element of this system is the smart meter which is the combination of all energy metering and intelligence. Smart meters are fundamentally different from ordinary electromechanical meters. Nevertheless the electromechanical meters that we use today have many drawbacks like poor accuracy and lack of configurability. They have many moving parts that are prone to wear out over time with varying operating temperature and conditions. They provide only energy measurement data and these data do not help in promoting the energy efficiency. Gathered data are inherently limited and the cost is high due to man power requirements. Considering these factors, smart meters are introduced for better energy saving, demand management and energy efficiency. Smart meters are capable of communicating with each other and executing command signals remotely and locally. They provide good solutions in overcoming the problems that are faced with the old grid system [11].

Figure 1.3 shows a conventional metering system. It doesn't provide a two-way interaction between the power utility and the consumer, since most of the meters are read monthly. Gathered data are limited and cannot be used for demand response [12].

Data collected from smart meters are important to multiple parties including consumers, energy utilities, and marketers. The consumers can use the data to adjust their load and save the electricity bill. The utilities can use these data to monitor the electricity usage of each consumer, to analyze the demand and to determine the electricity bills. Marketers can use the data to profile customers for specific advertisements [13].

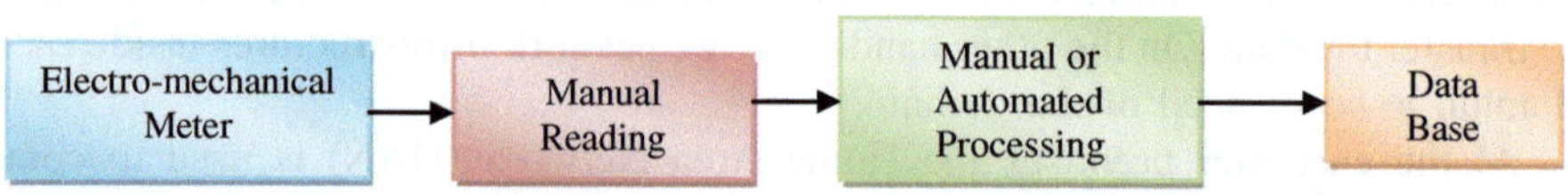

Fig. 1.3 A conventional metering system

Many advantages are attributed to smart metering, including lower metering cost, energy savings for residential customers, more reliability of supply, and variable pricing schemes to attract new customers. Smart meters can be used at the consumer's premises in order to monitor and control the home appliances and devices. They are capable of collecting diagnostic information about the distribution grid, and consumer appliances. The smart grid system is also supported with smart meters by measuring electricity consumption, supporting decentralized generation sources and billing the consumer remotely [11].

Smart meters are embedded with different technologies and services. Therefore implementation of smart meters appears to be multi-standard and incompatible. Smart meters are still evolving and many governments, organizations and companies are trying to establish different standards and policies [14]. Regardless the standards or policies, the meter developers ought to include the consideration of basic abilities such as:

1. Remote provision of metering data and related information to the utility
2. Two-way communications between the meter and the utility
3. Remote operation for disabling and enabling supply
4. Provision of information to home and networks
5. Load management at the consumer side
6. Exported electricity measurement to support DG such as PV and wind
7. Security tamper detections and remote configurations.

Figure 1.4 shows a typical smart metering system which includes smart appliances, auxiliary switches and generation. The HAN provides communication between meters, communication hubs, in-home display (IHD) units, and load control devices within the premises. The WAN is used for communication between the premises and the control center [15].

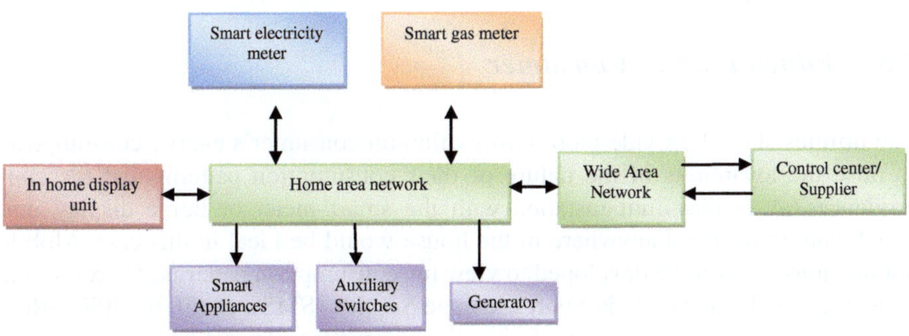

Fig. 1.4 A typical smart metering system

1.4 Energy Saving Through Smart Metering

A large number of Time of Use rates can be supported with smart meters. Therefore the utilities can set different tariffs at different times of the day in which the true cost of generating electricity is reflected [16]. They can also offer premium rates for occasional peaks in demand which is known as critical peak pricing. These methods are very effective for improving the pattern of power consumption with smart meters [17]. Consumers can be updated with warning signals before peak rates are applied [16]. For an example consumers can avoid using high power equipments (washing machines, water pumps, irons and refrigerators) at higher tariff levels or peak demands. When the smart meter has access to home appliances like refrigerators or air conditions, it can set the temperature value below the reference value at off peak demands. Therefore at peak hours the operation time can be reduced to save the cost while keeping the temperature at the desired level. On the other hand when the consumer has an electric vehicle, he can charge the battery at off peak hours and provide energy back to the system at peak hours. Sometimes the consumer may have a net metering system. Then he can sell the energy back to the system at peak demands to earn higher credits.

The energy management system is another system which promotes energy savings. This system doesn't require change in tariff, but induces self power savings by offering detailed information of energy consumption with time [17]. They are equipped with smart meters, home display units, and communication systems. The consumer can receive various information of power consumption which can be displayed in a home display unit. This information will help to analyze power consumption patterns and thereby reduce the consumption [17].

However this kind of energy saving can be achieved by proper demand side management. Otherwise nothing will change by replacing the old meters with smart meters, instead they are only digital meters that would transmit and receive data.

1.5 Techniques Used for Demand Side Management

1.5.1 Educating the Consumer

The utilities should provide more information on consumer's energy consumption to raise the awareness of the nature of their consumption patterns and the consequences. A display unit installed with the smart meter or home display unit which can be installed anywhere in the house would be ideal in this case. Mobile applications can also be developed to view his consumption patterns. Therefore the consumer can be updated via Short Message Service (SMS), email, or GPRS. Real time information about the current energy consumption, its cost, real time pricing, estimated cost for the month, credits earned by exporting energy to the system, and indication of any power cuts should be provided to the consumer. This kind of information will help consumers to change their behavior to achieve absolute reduction in energy use or shifting energy use from peak hours to off-peak hours. Cut down of unnecessary peaks of the system and exported energy from renewable sources will lead to a reduction in carbon dioxide emission. However, several researches point out that this is only effective where the consumer is already interested in this information. These types of consumers are susceptible to the messages of such data [18].

1.5.2 Incenting the Consumer

In this case, more sophisticated approach for tariffing is required than the typical flat tariff scheme. Time of day or time of use tariff structures can be introduced. The tariff schemes can be adjusted for the days of week and seasons accordingly. These kinds of schemes persuade the consumer to reduce the consumption at peak hours to save on electricity bill. Nevertheless weekends, public holidays and seasonal days could have different profiles. For an example the Ontario Energy Board in Canada uses this scheme [18].

The second method is real time pricing tariff. This tariff structure reflects the actual cost of power generation. In this method consumers are charged with prices that vary over short periods of time (typically hourly). Compared to other tariff methods, the real time tariff is more efficient. It can play an important role in a portfolio of strategies for cost-effectively meeting utility load obligations [19].

However these schemes require a change in measurement practice and require types of technologies that have been long present in the cellular communications market. A combination of a smart meter and intelligent meter data management system is also required [18].

1.5.3 Controlling the Consumer

Some consumers have agreements with the energy retailer to access and control their loads. Here the customer would allow the retailer to regulate their air-conditioner or refrigerator or space heaters in order to shed load at times of excessive demand. Significant reduction of overall load in a big city can be caused by reducing air-conditioning across all offices and homes. This methodology can be easily implemented with smart meters and smart appliance controllers [18]. Basically smart appliances monitor the power usage, protect the equipments, and automatically adjust its operation to the need of its owner. Some countries currently use these systems to monitor and control the loads remotely. On the other hand by using this technology, the electricity supply can be disconnected when credits are over especially for pre paid consumers [20].

1.6 Carbon Benefits Through Smart Metering

Smart meters promote energy efficiency and carbon reducing potential. Customers can be informed with the price and the carbon content of their fuel in order to reduce the electricity usage. This allows electricity generators to avoid having to keep big power plants warm to meet spikes in demand. Oil fired or gas fired generators are connected to the grid to match the peak demands. Generation cost of 1 kWh of these generators is very expensive and carbon emission is also high. Therefore if we can reduce the peak demands by using smart meters we don't need to run oil or gas fired generators at peaks. That will help us to reduce carbon emission to a significant level. On the other hand smart technologies which manage the end use demands can operate equipments and adjust building systems for top efficiency.

People are encouraged by many governments to adopt renewable energy in order to reduce the greenhouse gas emission. With the introduction of smart meters, consumers are more aware on using renewable energy and try to generate energy at their premises using wind or solar energy. They can use the generated energy for their own consumption and sell the additional energy back to the system to earn some money. Even though it is not cheap to install renewable power generating systems at the moment but with the development of technology it will be a cost effective method [18]. Small scale distributed energy may also include the micro turbines and fuel cells. This kind of local generation contributes to recycling waste heat into the building systems and to encourage more efficient usage of fuel than in central generating stations [21].

Cleaner power options offered on the market will be automatically sought by smart equipment, appliances and building controllers. These equipments are capable of coming online during air pollution alerts or operating at all times to express owner desires for green power options [21].

1.7 Smart Metering Projects

Most countries in the world are ready to take a step with smart grid. There are plenty of main projects and pilot projects on smart metering which are conducted in several countries in order to achieve energy efficiency and conservation. Following are few examples of them.

1.7.1 Smart Metering Projects in Italy

Before smart metering projects were started, Italy had experienced high cost of energy consumption than other European countries. Electricity prices were also higher. A system was needed for better energy conservation as well as fraud detection to minimize the electricity cost and carbon emissions [22]. The state owned utility, “ENEL”, installed 30 million smart meters over a 5 year period, 2000–2005. The total cost was about 70 Euros per house. The savings to ENEL are 500 million Euros per year and its total investment will be paid back by the end of the decade. Total power consumption from households has fallen 5 % as a result of the switch (as opposed to rising about 1.5 % a year in the UK). They have provided many features with smart meters including ability to turn power on and off at demand side, remote billing, service outage detection, fraud detection and multi-tariff schemes. And also new plans have been introduced for prepaid customers [23]. They have used standards-based power line technology from Echelon Corporation to transmit the data to a central office [22]. Finally, Italy has showed a good example to the world on energy saving via smart metering.

1.7.2 Smart Metering Projects in USA

California is the main state in USA where smart meters are used in large scale. They have experienced a summer peak demand for power during about 50–100 hours per year. The main reason for this was the high energy consumption for air-conditioning. An efficient system was needed to reduce this demand to some extent. California’s energy regulators approved a program to install nine million smart meters in the Northern California territory of Pacific Gas and Electric (PG&E) company in 2006. Using these meters they were able to receive electric consumption in each household on an hourly basis. This enables PG&E to set pricing that varies by season and time of the day, rewarding customers who shift energy use to off-peak periods [23].

California Public Utilities commission has approved a $1.7 billion smart meter proposal of (PG&E) Company. This has allowed the utility to move forward with major investment in new smart metering to achieve energy efficiency and conservation. Fast response to power outages can also be achieved. Restoration services can be implemented faster due to messages from consumer services. Customers can monitor their consumption patterns online and can make better decision to manage their usage. They are well aware of the tariff schemes and taking actions to save on their bill by shifting their energy use from peak hours to off-peak hours. A lot of companies in USA are willing to use smart meters. However the largest program in USA reported in 2010 is the PG&E’s program [22].

The Los Angeles Department of Water and Power which is the largest municipal utility in USA has decided to expand their AMI. The utilities commercial and industrial customers are encouraged by AMI to save on their bill. They can adjust their consumption levels with the data provided by smart meters. Ultimately this has caused a boarder reduction on monthly electricity bill as well as carbon emissions [23].

Austin Energy which facilitates electricity to 400,000 customers began deploying a two-way RF mesh network in 2008. The estimated coverage was 260,000 residential smart meters. Oncor Electric Delivery uses advanced metering system which supports 15-min-interval data, remote disconnections and a home area network. This system provides text messages, pricing signals and load control to home users through smart meters [23].

1.7.3 Smart Metering Projects in Asia Pacific

In Asia Pacific, utilities are beginning to develop and install smart meters. According to a new report from Pike research, the total installed base of smart meters in the region will see a compound annual growth rate of 37%, increasing from 52.8 million in 2010 to over 350 million by 2016 [23].

1.7.4 Smart Metering Projects in Canada

Ontario smart meter project has been identified as a successful project in Canada. People with smart meters are charged a sliding scale of rates based on the time of day. Prices range from a high of 9.3 cents per kWh to a low of 2.9 cents per kWh. It would cost much less to use major appliances after 10 p.m. In the summer, it would make sense to use the air conditioner less during the afternoon. Holding off until the weekend for laundry would also save money. The Ontario Energy Board said that costs would be recovered from consumers over time. The board expects it to cost more than $1 billion to install the meters across the province. That should add between 1 and 4 dollars a month to the average electricity bill [23].

1.7.5 Smart Metering Projects in UK

In the United Kingdom most of the conventional meters have been replaced by smart ones. This is considered as the largest program ever undertaken which involves 27 million domestic customers. More than 40 % of the benefits are identified in business case where the savings are achieved through the time of use tariffs. Proponents said that there would be benefits to the network operators in terms of carbon savings [23].

1.7.6 Smart Metering Projects in China

Currently, China is investing in large scale for expansion of the national energy infrastructure to match the rapid increase in demand [24]. This is because China has a very rapid economic growth and rapid development of their technologies. On the other hand, they have an uneven geographical distribution of electricity generation and consumption [25]. The State Grid Corporation of China has been involved in smart grid construction. The company purchased 70 % of 17.4 million smart meter units shipped worldwide in the first quarter of 2011. According to government targets, China is going to produce over 300 million smart meters in the next 5 years. Most of China's power (around 70 %) is generated by dirty coal plants, hence the government has started to shift the power demands to renewable sources.

They target to achieve 15 % of the total power supply by using renewable sources in 2020. The targeted resources are wind power and solar [22]. The State Grid Corporation supports the establishment of a national wind and solar energy research center. The carbon dioxide emissions will be reduced by 10.5 billion tones over the next 10 years. The government has declared that the Carbon emission per-unit of gross domestic product (GDP) will reduce 40–50 % than 2008 by the year of 2020. The government has facilitated the development of electric cars as well. The old grid will be replaced with smart grid [25]. The first green city in the world is also under construction in China.

1.7.7 Smart Metering Projects in Japan

Japan has already stepped to smart grid which involves and engages communities across the state. Now Japan has become an ideal partner for smart grid development and has received an international recognition. Economic development, reduction in energy cost, improvement in electric grid reliability, and more environmental quality will be achieved through smart grid [26].

In 2009, Japanese government explained that they would reduce the carbon emissions by 75 % of those in 1990 or two-third of those in 2005. They have to supply 28 GW and 53 GW of photovoltaic power to the grid by 2020 and 2030 to reduce the carbon emissions. Therefore, three study committees have been appointed since 2008 by the Ministry of Economy, Trade and Industry [1].

After the Tsunami and the Fukushima disaster, Japan faced with rolling blackouts and shortage of supply. Thereafter the Tokyo Electric Power Company started to supply smart meters to 10 million customers. Large-scale grids will be developed in future which is capable of handling power like the internet handles the data [27].

References

1. Ekanayaka J, Liyanage K, Jianzhong W, Yokoyama A, Jenkins N (2012) The smart grid. In: Smart grid technology and applications, 1st edn. Wiley, UK, pp 1–15
2. Smart Grid (2012) http://en.wikipedia.org/wiki/Smart_grid. Accessed 2 June 2012
3. Oksman V (2010) Use of G.hn transceivers for smart grid and EV applications version 1.0. In: IEEE P2030 TF3. Available at https://mentor.ieee.org/.../2030-10-0042-00-0004-use-of-g-hn-transc. Accessed 25 feb 2013
4. U.S. Department of Energy, Smart Grid System Report (2009) http://www.oe.energy.gov/sites/prod/files/oeprod/DocumentsandMedia/SGSRMain_090707_lowres.pdf. Accessed Jan 2012
5. Department of environment, water, heritage and the art of Australian government, smart grid, smart city grant guidelines (2013) www.ret.gov.au/.../smart-grid/smartgrid-grant-guidelines-doc.doc. Accessed 25 Feb 2013
6. U.S. Department of Energy, Smart Grid (2012). http://www.smartgrid.gov/the_smart_grid. Accessed 3 June 2012
7. Eichhorn J (2010) Considerations in building a smart grid communications network. http://www.smartgridresearchconsortium.org/Eichhorn.ppt. Accessed 25 Feb 2013
8. Saremi F, PoLiang Wu, Heechul Yun (2013) Smart grid.http://www.courses.engr.illinois.edu/cs598tar/fa2010/ClassNotes/Grid.ppt. Accessed 25 Feb 2013
9. Canadian Electricity Association (2012) The smart grid: a pragmatic approach. www.electricity.ca/media/SmartGrid/SmartGridpaperEN.pdf. Accessed 2 June 2012
10. Smart Grid conceptual model (2013) In: Smart grid. IEEE. Available via http://smartgrid.ieee.org/ieee-smart-grid/smart-grid-conceptual-model. Accessed 26 Feb 2013
11. Depuru SSSR, Wang L, Devabhaktuni V, Gudi N (2011) Smart meters for power grid—challenges, issues, advantages and status. In: Proceedings IEEE power systems conference and exposition, Phoenix, 20–23 Mar 2011
12. Gerwen RV, Jaarsma S, Wilhite R (2006) Smart metering.www.idc-online.com/technical_references/.../Smart_Metering.pdf. Accessed 3 June 2012
13. Privacy on the smart grid (2013). In: IEEE spectrum. Available via http://spectrum.ieee.org/energy/the-smarter-grid/privacy-on-the-smart-grid. Accessed 26 Feb 2013
14. Xu FY, Zhou , Wu YL, Ma Y (2010) Standards, policies and case studies in smart metering. In: Proceedings IEEE power and energy society general meeting, Minneapolis, 25–29 July 2010
15. Coaster M (2010) Smart metering implementation program: statement of design requirements. www.ofgem.gov.uk/.../Smart%20metering%20. Accessed 3 June 2012
16. A Guide to Smart Metering: empowering people for a better environment (2012) www.esmig.eu/.../ESMIG%20- Accessed 3 June 2012
17. Choi TS, Ko KR, Park SC, Jang YS, Yoon YT, Im SK. (2009) Analysis of energy savings using smart metering system and in-home display. In: Proceedings IEEE transmission and distribution conference and exposition, Seoul, 26–30 Oct 2009
18. Smart Metering-Enable Greater energy Efficiency (2012) http://www.enterprise.alcatel-lucent.com/...docs/SmartMetering_wp_092107.pdf. Accessed 5 June 2012
19. Barbose G, Goldman C, Neenan B (2004) A Survey of utility experience with real time pricing. http://www.eetd.lbl.gov/ea/EMP/reports/54238.pdf. Accessed 5 June 2012
20. Smart Grid Insights: Smart Appliances (2010) www.zpryme.com/SmartGridInsights/2010_Smart_Appliance_Report. Accessed 5 June 2012
21. Mazza P (2012) A northwest initiative for job creation, energy security, and clean affordable electricity. In: Power up the smart grid.www.wutc.wa.gov/.../powering%20up%20smart%20grid%20report. Accessed 10 June 2012
22. Zhou L, Xu FY, Ma YN (2010) Impact of smart metering on energy efficiency. In: Proceedings machine learning and cybernetics international conference, Qingdao, 11–14 July 2010

23. Smart Metering (2012) http://www.en.wikipedia.org/wiki/Smart_meter. Accessed 12 June 2012
24. Ryberg T (2012) Smart metering in North America and Asia-Pacific. www.berginsight.com/ReportPDF/ProductSheet/bi-smnaap-ps.pdf. Accessed 13 June 2012
25. State Grid Corporation (2012) http://www.internationalrivers.org/campaigns/state-grid-corporation. Accessed 14 June 2012
26. Smart Grid Japan (2012) http://www.smartgridjapan.com. Accessed 14 June 2012
27. Japan Jumps on the Smart Grid Bandwagon (2012) http://www.smartgridnews.com. Accessed 15 June 2012

Chapter 2
Evolution of Electricity Meters

Abstract This chapter describes the evolution of electricity meters from electromechanical meters to modern smart meters. The operation principle of a typical electromechanical meter is well described with illustrations and calculations. The drawbacks of electromechanical meters are also highlighted. The operations of reactive energy meter and maximum demand meter are also discussed. The technological evolution from solid state electronic meters to smart meters is discussed. The importance of smart meters in modern energy measurement is highlighted. The basic hardware structure of a modern smart meter is illustrated. A broader discussion is done about the hardware components inside a smart meter including voltage and current sensors, power supplies, energy measurement, microcontroller, real time clock and communications. Importance of standardization of smart meters is described.

2.1 Introduction

In recent years, domestic and industrial users have shifted from traditional meters to smart meters. Electromechanical meters were a dominant part of electricity measurement before 1970 [1]. They could only measure the electrical energy. However it had been identified that the requirement of a meter which could communicate and measure the electrical energy along with other electrical parameters. Therefore solid state electronic meters were introduced to measure the overall electrical parameters.

Between 1970 and 2000, automatic meter reading was added to electronic meters and it was a great achievement since it could send the data in near time. However it could only provide the one-way communication. This limitation was overcome by the introduction of smart meters which can provide two-way communication. Smart meters can measure all the electrical parameters like electronic meters and communicate data in a meaningful way. The consumer is updated with

K. S. K. Weranga et al., *Smart Metering Design and Applications*,
SpringerBriefs in Applied Sciences and Technology,
DOI: 10.1007/978-981-4451-82-6_2,

electricity usage, cost, tariffs and other notifications sent by the utility. Smart meters have different functionality to manage the end user loads and run them in an optimal way to reduce the electricity bill as well as to conserve the energy. Smart meters have been used since a decade [1].

Nevertheless many researchers and developers are trying to add features to smart meters and try to come up with best solutions for energy efficiency, conservation and demand management. Smart meters are still evolving and many governments and organizations are trying to standardize them.

2.2 Operation Principles of an Electromechanical Energy Meter

Electromechanical energy meter is the most traditional and widely used energy meter over a century. It is capable of measuring only the active energy which is typically displayed on a mechanical counter in kWh. Figure 2.1 shows an example of a typical single phase electromechanical meter [2].

It is basically designed with four major systems which are driving system, moving system, breaking system, and registering system. The driving system consists of two electromagnets while the moving system consists of an aluminum disc. The permanent magnet acts as the breaking system while the gear train and counter act as the registering system. The electromagnetic force is produced by the

Fig. 2.1 A single phase electromechanical meter

arrangement of voltage and current coils. The voltage coil is connected across the supply while the current coil is connected in series with the load [3]. The voltage coil produces a magnetic flux in proportion to the voltage and the current coil produces a magnetic flux proportional to the current. The aluminum disc is mounted on a rigid axis. A mechanical force is exerted on the disc by the Eddy currents produced. The register mechanism integrates the speed of the disk over the time by counting the number of revolutions [4]. Figure 2.2 shows the basic arrangement of a single phase electromechanical energy meter.

Current coil or the series coil produces alternating flux which is proportional and in phase with the load current. Voltage coil or the shunt coil carries a current proportional to the supply voltage. The flux produced by the voltage coil is not in phase with the supply voltage. This flux is 90° lagging with the supply voltage. This is done by having properly adjusted copper rings in the flux path as shown in the Fig. 2.2. However some electromechanical meters use winding with the series connected lag adjusting resistor to perform this task [4].

The phasor diagram of the single phase energy meter is shown in Fig. 2.3 [5].

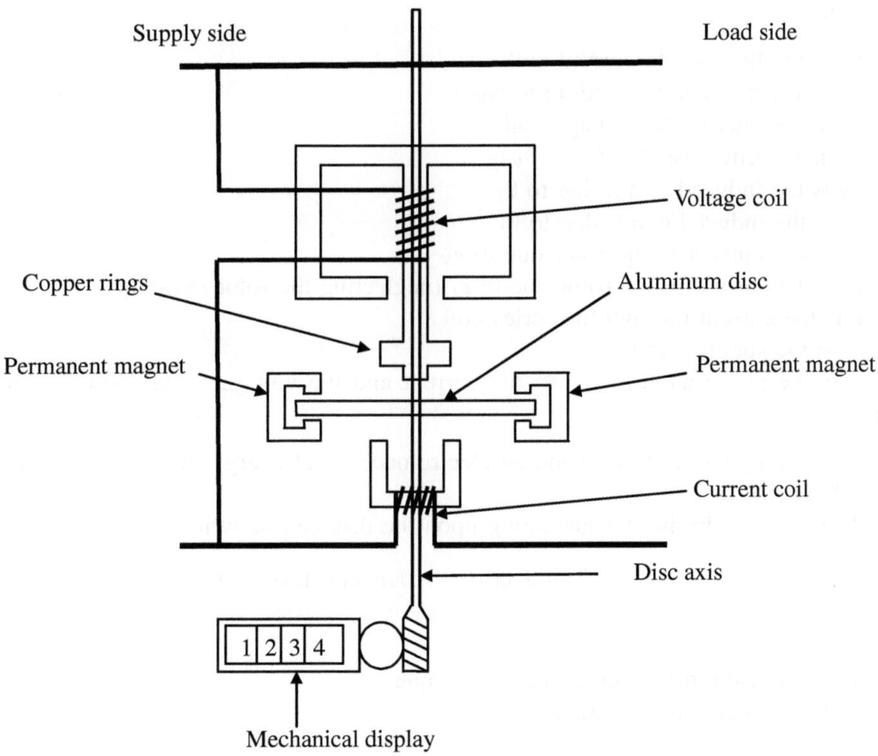

Fig. 2.2 Components of a single phase electromechanical energy meter

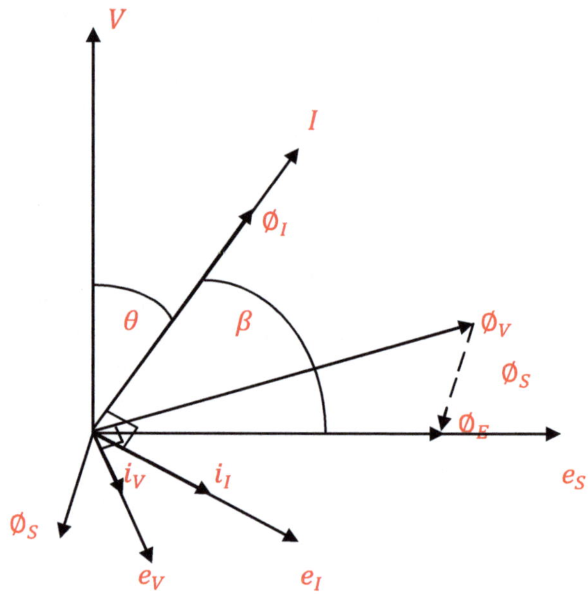

Fig. 2.3 The pahsor diagram of the single phase energy meter

where;

$\emptyset_I$ is the flux due to current coil
$\emptyset_S$ is the flux due to quadrature band
$\emptyset_V$ is the flux of the voltage coil
$\emptyset_E$ is the effective flux $(\emptyset_S + \emptyset_V)$
e_S is the induced e.m.f. due to $\emptyset_E$
e_I is the induced e.m.f. due to $\emptyset_I$
i_V is the current in the rotor due to e_V
i_I is the current in the rotor due to e_I (neglecting the rotor resistance)
I is the current through the series coil
V is the supply voltage
θ is the phase angle between the current and the voltage (load connected to meter)

The energy calculation inside an electromechanical energy meter is described as follows [6].

The average driving torque acting upon the disc can be written as

$$T_{d(av)} = k_d[\emptyset_E i_I \cos(\theta) - \emptyset_I i_V \cos(180 - \theta)] \tag{2.1}$$

where;

$T_{d(av)}$ is referred to average driving torque
k_d is a constant for the meter

Since

$$\emptyset_E \propto \frac{V}{\omega} \rightarrow \emptyset_E = k_1 \frac{V}{\omega} \tag{2.2}$$

$$\emptyset_I \propto I \rightarrow \emptyset_I = k_2 I \tag{2.3}$$

$$i_I = k_2 \frac{I}{Z} \omega \tag{2.4}$$

$$i_V = k_1 \frac{V}{Z} \tag{2.5}$$

where; Z is the eddy current path impedance (the phase angle is assumed to be zero), k_1 and k_2 are constants, ω is the angular frequency.

Substituting terms in (2.1), using (2.2)–(2.5)

$$T_{d(av)} = k_d \left[k_1 \frac{V}{\omega} k_2 \frac{I\omega}{Z} \cos(\theta) - k_2 I k_1 \frac{V}{Z} \cos(180 - \theta) \right] \tag{2.6}$$

$$\begin{aligned} T_{d(av)} &= \frac{k_d k_1 k_2}{Z} [VI\cos(\theta) + VI\cos(\theta)] \\ T_{d(av)} &= \frac{2k_d k_1 k_2}{Z} VI\cos(\theta) \\ T_{d(av)} &= k' P \end{aligned} \tag{2.7}$$

where; $k' = \frac{2k_d k_1 k_2}{Z}$ and P is the active power

Form (2.7), we can see that the driving torque is directly proportional to the active power.

The breaking torque is produced by two permanent magnets mounted in opposite directions.

$$T_{b(av)} = k_b \emptyset_b i_b \tag{2.8}$$

$$\begin{aligned} T_{b(av)} &= k_b \emptyset_b \frac{e_b}{R_e} \\ T_{b(av)} &= k_b \emptyset_b \frac{N\emptyset_b}{R_e} \\ T_{b(av)} &= \frac{k_b \emptyset_b^2}{R_e} N \end{aligned} \tag{2.9}$$

where;

$T_{b(av)}$ is referred to the average breaking torque
i_b is the eddy current due to $\emptyset_b$
e_b is the induced e.m.f. due $\emptyset_b$
k_b is a constant of proportionality,
$\emptyset_b$ is the flux produced by permanent magnets,
N is the rotational speed of the aluminum disc and
R_e is the resistant of the eddy current path.

At steady state

$$T_{d(av)} = T_{b(av)} \tag{2.10}$$

Therefore using (2.7) and (2.9);

$$k'P = \frac{k_b \emptyset_b^2}{R_e} N$$

$$P = \frac{k_b \emptyset_b^2}{k' R_e} N \tag{2.11}$$

From (2.11) we can clearly see that the active power is proportional to the rotational speed of the disc.

The disk axis will transmit the disc rotation to a mechanical counter. This system continuously accumulates the disc displacement with the time. Therefore

$$\alpha = \int N dt \tag{2.12}$$

$$E = \int P dt \tag{2.13}$$

where;

α is the disc displacement
E is the active energy and
t is the time

Using (2.11)–(2.13)

$$E = \int \frac{k_b \emptyset_b^2}{k' R_e} N dt \tag{2.14}$$

$$E = \frac{k_b \emptyset_b^2}{k' R_e} \int N dt$$
$$E = \frac{k_b \emptyset_b^2}{k' R_e} \alpha \tag{2.15}$$

According to (2.15), the displacement is proportional to the active energy. Therefore the counting system can be calibrated accordingly with the active energy to display the energy consumed.

2.3 Drawbacks of the Electromechanical Energy Meters

Electromechanical meters react to the changes more slowly than digital meters. They have many susceptible errors due to environmental variations and regular operations [7]. The moving parts inside these meters are prone to wear over time, varying temperature, and conditions. On the other hand mechanical gears wear due to effects of dirt, dust and humidity. The gear ratios also change over time due to lack of lubricants. Nevertheless, vibration and shock affect the accuracy of the meter in the long run. Therefore periodic calibrations are required at regular intervals. Furthermore, due to the lack of linearity of iron core and the inertia of the spinning disk, errors can be caused at low and high loading [7].

Electromechanical meters require manual readings. In other words meter readers have to go and take the reading manually to issue the bill. Because of the man power requirement, there is always an additional cost to the bill apart from the energy consumed. Moreover the tampering of meter readings, human errors in readings, irregularities in billing time, and controversial billing are also possible with manual readings [8].

Power theft is a major problem caused with electromechanical meters. Illegal reconnection of power lines, bypassing of the energy meter, and very weak conditional access enforcement cannot be detected directly with these meters [9].

2.4 Reactive Energy Meters

The active energy doesn't represent the total energy delivered to a consumer. Therefore the measurement of reactive energy is also an interesting part to analyze the overall energy delivery. The reactive energy had been traditionally measured by reactive energy meters before the digital meters were introduced. Reactive energy meters are also known as sine meters because they measure the reactive component of the current which is 90° apart from the applied voltage. Even though the appearance is same as the typical electromechanical meter the construction of the sine meter is little bit different. The voltage coil is used to produce voltage flux which is in phase with the supply voltage. This is done by adding a high non-inductive resistor in series with the voltage coil. Meanwhile the current coil produces a flux in phase with the load current. The torque is proportional to the product of volts $\times$ amperes $\times$ $\sin\theta$, where θ is the phase angle between the voltage and current. This meter reads the reactive energy typically in kvarh. A register mechanism is used to count the units of reactive energy. The lagging currents cause forward count in registers while leading currents cause backward count. The accuracy of measurement is lower than electromechanical meters due to some limitations in design [5]. The phasor diagram of a single phase sine meter is shown in Fig. 2.4.

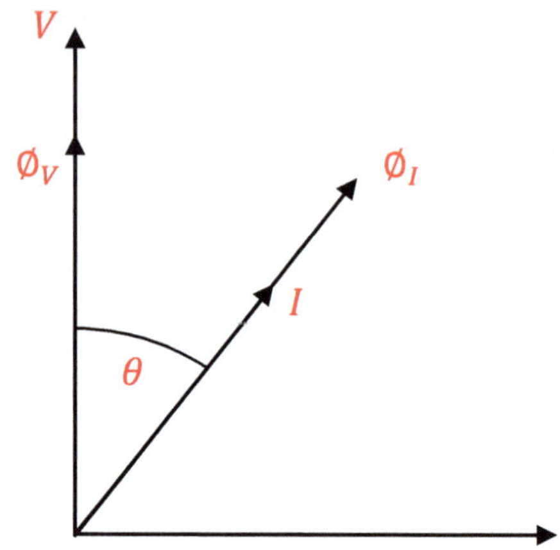

Fig. 2.4 The phasor diagram of a single phase sine meter

2.5 Maximum Demand Meters

Maximum demand is also taken into consideration for bulk consumers in electricity billing. Even though the real mechanical work is done due to the active energy delivered to the consumer, additional cost is paid for the maximum demand. This cost is designed to encourage the consumers to operate near to unity power factor. Maximum demand is measured with maximum demand meters. There are three main types of maximum demand meters [5]. They are

- Integrated instruments
- Thermal type indicators
- Electromagnetic type meters.

Different types of meters have different arrangements and operation mechanisms. However the maximum demand calculation is common to every meter. Maximum demand is the highest average demand recorded over specified time intervals. The average demand within a period of time can be written as

$$MD_i = \frac{1}{T}\int_0^T f(t)dt \tag{2.16}$$

where;

T is the demand interval or the demand integrating period
MD_i is the average demand over the time interval T
$f(t)$ is the demand function (in kW or kVA)

The demand interval(T) may vary power utility to other. Typical values for T are 10, 15, and 20 min. The maximum demand is the highest value among all the average demands recorded over the month. Therefore the maximum demand is

$$MD_{month} = Maximum(MD_i) \tag{2.17}$$

2.6 Electronic Meters

Electronic meters are capable of measuring electricity usage with digital technology. At the same time they can measure the other electrical parameters such as phase voltages, phase currents, frequency, power factor, active power, reactive power, apparent power, maximum demand, and power quality measurements. Therefore they perform all the tasks that are done by the other types of meters. They have also the capability of sending the measured data through a communication link.

A Typical electronic meter consists of a power supply, microcontroller, Real Time Clock (RTC), LCD display, and communication ports [4]. It has voltage

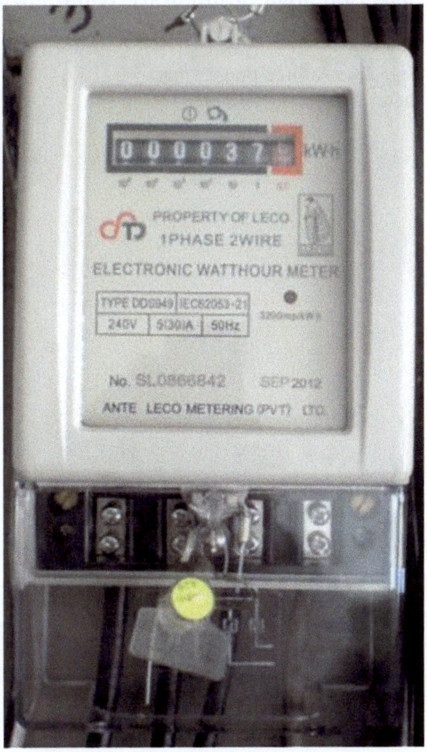

Fig. 2.5 A single phase electronic energy meter

inputs, current inputs and a reference voltage. Voltage and current signals are processed to measure and display the electrical parameters.

Electronic meters provide timely data, high accuracy in measurement in a wide range of loads, greater flexibility of design, and updating capacity. They are not influenced by external magnets or orientation of the meter itself. Therefore digital meters are more reliable and tamperproof than electromechanical meters [2]. A single phase electronic energy meter used by a utility company is shown in Fig. 2.5.

2.7 Smart Meters

Smart meters are different from electronic meters because of their additional functionalities and features. Apart from electricity measurements and automatic meter reading (AMR), they allow two-way communication between the meter and the base station. Load profiling, pre-payment, remote disconnection and reconnection, power outage notification, tamper detection, and multi-tariffing are also possible with smart meters [10]. A three phase smart meter is shown in Fig. 2.6.

The electronic meters have been used effectively for accurate billing. However more functions are needed such as remote readings, outage detection, tamper detection, load profiling for better customer service and reliable supply. Therefore

Fig. 2.6 A three phase smart meter

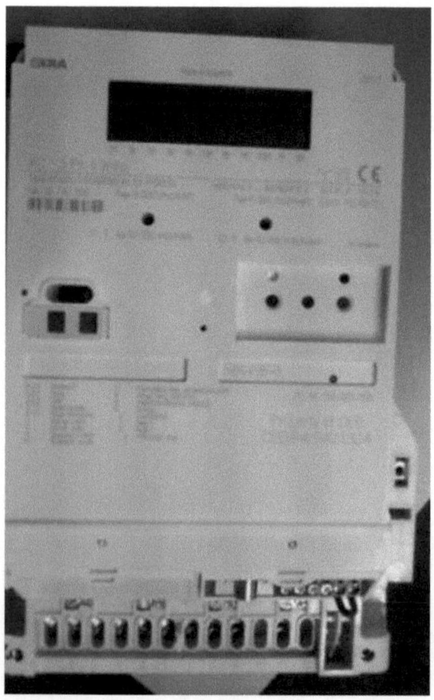

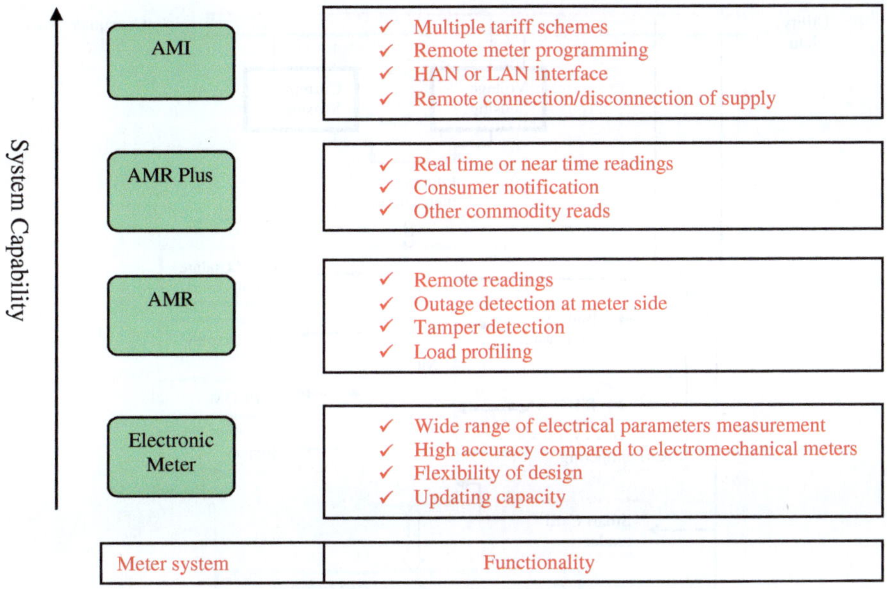

Fig. 2.7 Smart meter technology evolution

AMR system is introduced by combining the communication infrastructure to electronic meters. Meanwhile more features and functions are added to AMR system. Ultimately AMI (Advanced Metering Infrastructure) has been developed to today's technology with two-way communication and data management system. Figure 2.7 shows the evolution of smart meter technology from electronic meters to AMI [11].

2.8 The Hardware Structure of a Smart Meter

Figure 2.8 shows the functional bock diagram of a smart meter. It includes signal acquisition, signal conditioning, Analogue to Digital Conversation (ADC), computation and communication [1].

Smart meters use voltage and current sensors to get the input signals. Signal conditioning, ADC, and computations are done inside the micro controller unit (MCU). Additional hardware components are required for other operations like

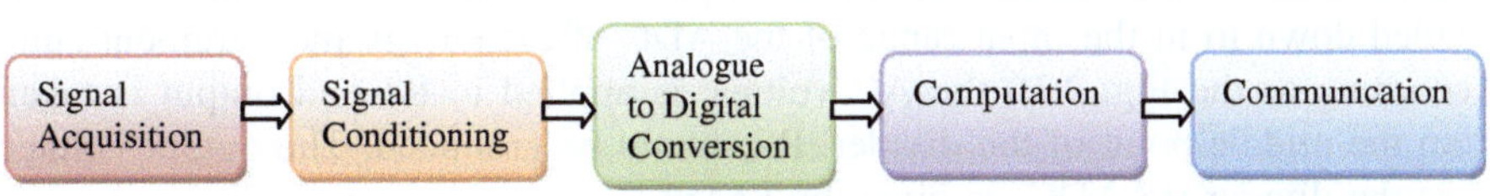

Fig. 2.8 Functional block diagram of a smart meter

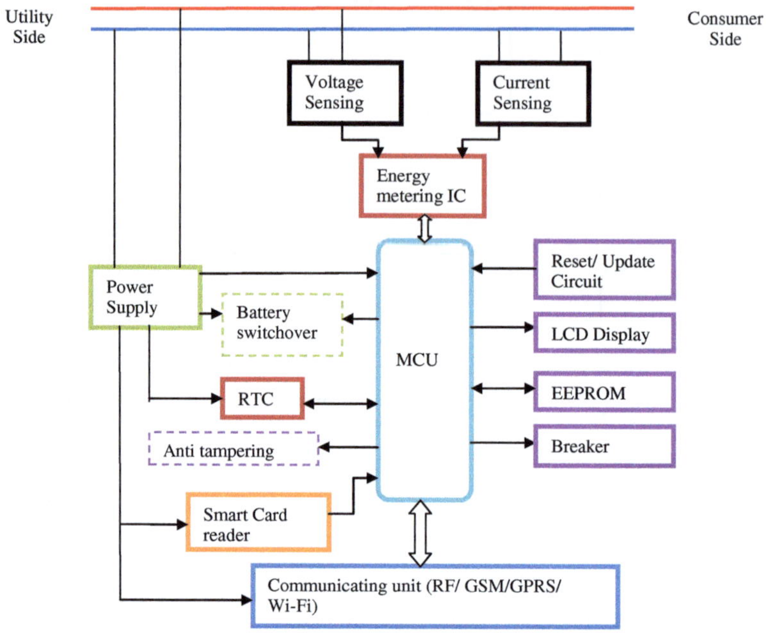

Fig. 2.9 Hardware structure of a modern smart meter

communication, time and date measurements, and data backup and storage. A smart meter is typically composed of following hardware components:

- Voltage and current sensing unit
- Power supply
- Energy measurement unit (metering IC)
- Microcontroller
- Real time cock
- Communicating system.

Figure 2.9 shows the hardware structure of a modern smart meter.

2.8.1 The Voltage Sensing Unit

Simple resistor dividers are widely used as voltage sensors in digital meters due to low cost. Figure 2.10 shows the configuration of a resistor divider type voltage sensor.

The values of R1 and R2 should be chosen such that the AC mains voltage is divided down to fit the input range of the ADC of the energy measurement chip. According to the Fig. 2.10 the AC voltage is applied to R1 and output is taken from the middle point of the divider. R2 should be grounded. The output voltage from the divider (to ADC) is given by (2.18).

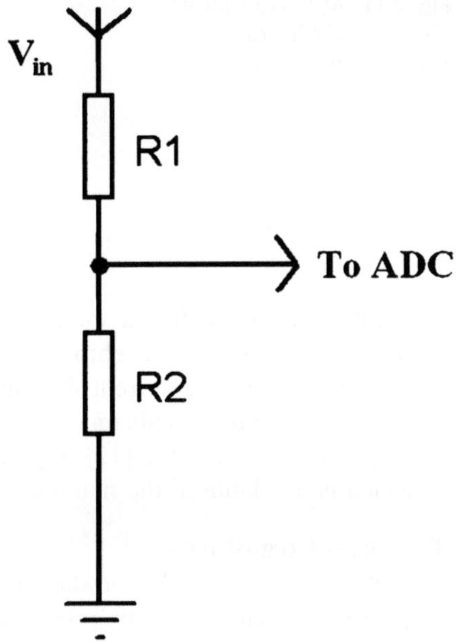

Fig. 2.10 Resistor divider configuration

$$V_o = \frac{R2}{R1 + R2} V_{in} \tag{2.18}$$

where

V_o is the output voltage
V_{in} is the input voltage

Normally R1 and R2 are in kΩ scale. R1 is much greater than R2 (R1 ≥ 500R2). Higher values of resisters are chosen because of the lesser power dissipation.

2.8.2 The Current Sensing Unit

The current sensing unit typically consists with current sensors and anti aliasing filters. Four types of current sensors are widely used in smart meters. They are:

- Hall effect-based linear current sensors
- Current transformers
- Shunt Resistor
- Rogowski coils.

Fig. 2.11 ACS 712 Hall IC (Courtesy of Allegro MicroSystems, Inc.)

(A) Hall effect-based linear current sensors

These sensors consist of a chip and a copper conduction path located near the surface of the die. The current flowing through the copper conduction path generates a magnetic field. This magnetic field is sensed by the Hall IC and converted into a proportional voltage [12]. Figure 2.11 shows an example of ACS 712 Hall IC which is available in the market.

(B) Current transformers

Current transformers (CTs) produce a secondary current which is proportional to the primary current. Magnetic properties of CTs are highly linear over wide range of primary current, and temperature. The primary is connected in series with the device [13]. The isolation is provided from primary to secondary side thus ensuring high reliability for metering devices. However the linearity depends on the magnitude of the primary current and the impedance of the secondary. Every CT is classified according to its performance. Normally class 0.1, 0.2, 0.5 and 1 are used for metering purposes. Although CTs are expensive than shunt resistors, they consume lesser power. However CTs have nonlinear phase response at low currents and large power factors [1]. Figure 2.12 shows the typical through-hole type current transformers which are widely used in metering applications [14].

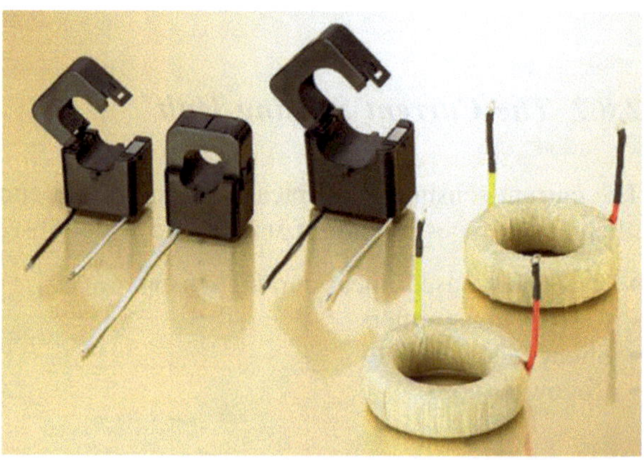

Fig. 2.12 Through-hole type current transformers (Courtesy of Taehwatrans Co. Ltd.)

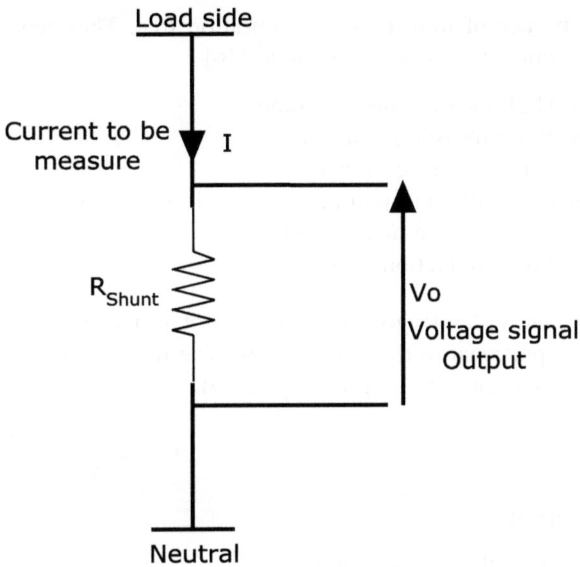

Fig. 2.13 Circuit diagram of a shunt resistor

(C) Shunt resistors

Shunt resistors are widely used in metering applications because of their lower cost than other types of current sensors. These sensors are simply placed in series with the load current path. Their resistances are typically in the range of 100 μΩ–500 mΩ. The power dissipated is proportional to the square of the current. Therefore a very small resistance should be selected to minimize the heat dissipation [1, 15]. Shunt resistors are highly stable resistors designed with low resisting materials so that the resistance doesn't change with the current, temperature or age. The voltage across the shunt resistor is proportional to the current that flows through it. This voltage signal is fed to an energy measuring chip or to a MCU. Therefore, when the resistance is known, the current can be calculated according to Ohm's law. The circuit diagram of a shunt resistor is shown in Fig. 2.13. Although resistive shunts are inexpensive, highly linear, and immune to magnetic influences, they do not have the inherent electrical isolation.

(D) Rogowski coil sensors

Rogowsli coils were introduced to the electrical industry as far back as in 1912, to measure the magnetic fields, but these could not be used for current measurements, since the power produced was not sufficient to drive electromechanical equipments. With the development of solid state electronics and microprocessor based systems, Rogowski coils have provided wide range of opportunities. These sensors are coils with non-magnetic core for which the name air cored is used. They give an output voltage which is proportional to the rate of change of current. They linearly convert the primary current up to all short circuit levels. Due to the absence of iron, they are saturation free. They have many advantages over conventional CTs, which include [16]

- High measurement accuracy
- Wide measurement range
- Wide frequency range
- Can withstand unlimited short circuit currents
- Small in size and weight
- Low production cost.

However Rogowski coils cannot produce a voltage signal which is directly proportional to the current flow. The relationship between the output voltage and the current flow is given by (2.19).

$$V = k\frac{dI}{dt} \tag{2.19}$$

where;

V is the output voltage
I is the current to be measured
t is the time and
k is a constant

Current signal should be recovered from dI/dt signal. Rearranging the terms in (2.19)

$$\begin{aligned} Vdt &= kdI \\ \int Vdt &= \int kdI \\ I &= \frac{1}{k}\int Vdt \end{aligned} \tag{2.20}$$

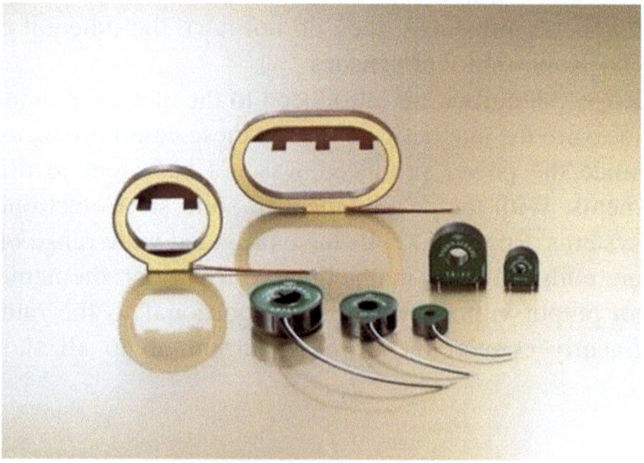

Fig. 2.14 Commercially available Rogowski coils (Courtesy of Taehwatrans Co. Ltd.)

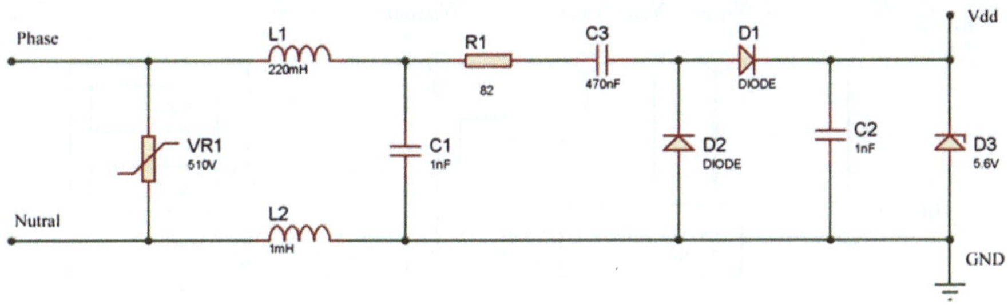

Fig. 2.15 Power supply schematic in STPM10 reference design (Courtesy of STMicroelectronics, Inc.)

According to (2.20), the output voltage should be integrated with respect to time to get the original current signal. Some energy measurement chips have built-in integrator to recover the current signal from Rogowski coils.

Rogowski coils are commercially available in a variety of configurations, including the popular flexible type. Figure 2.14 shows some of commercially available Rogowski coils [17].

2.8.3 Power Supply

Power supply unit may vary from smart meter designer to designer. However this unit typically consists of step-down transformers, rectifiers, AC-DC converters, DC–DC converters and regulators. Energy measurement chip designers provide their own reference power supply schematics. Figure 2.15 shows the power supply schematic which is used in STPM10 energy meter reference design [18].

However the power output from these circuits may not be sufficient to drive other hardware components in the smart meter. Therefore the required power to drive the energy chip, MCU, LCD, battery charger and communication unit should be taken into consideration before designing a power supply. Figure 2.16 shows the block diagram of a typical power supply used in smart meters. First the AC line voltage is rectified through a diode bridge. Sometimes the AC voltage is stepped down before the rectification. Then the unregulated voltage output is fed to a DC–DC converter or a regulator IC. DC–DC convertor consist of an inductor, a

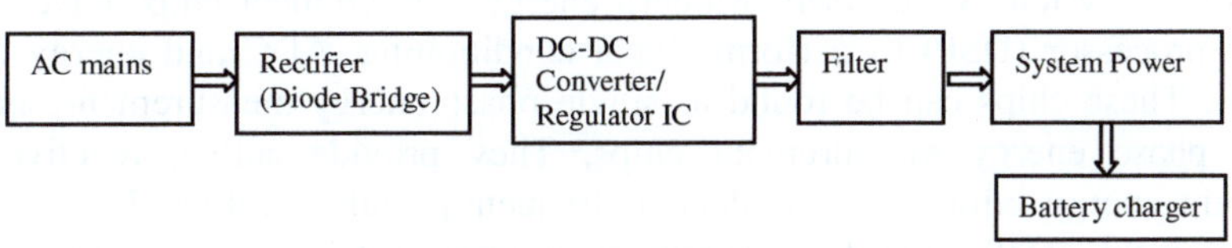

Fig. 2.16 A typical power supply used in smart meters

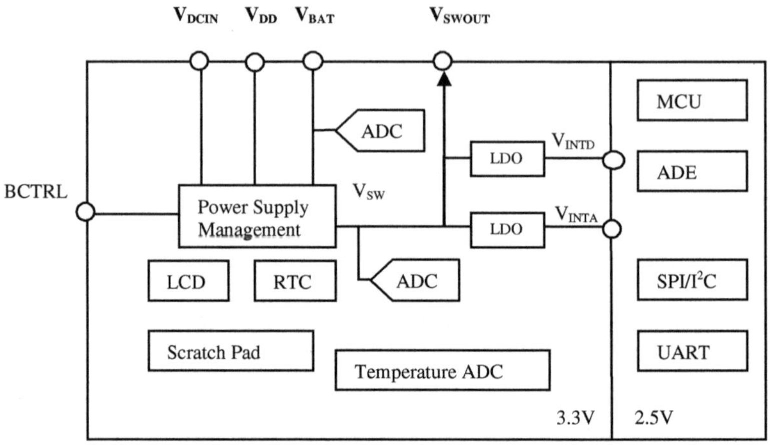

Fig. 2.17 Power supply architecture of ADE 5166 (Courtesy of Analog Devices, Inc.)

capacitor, and an electronic switch. The electronic switch in this converter can be a MOSFET, power transistor or IGBT. Else, a regulator IC is used instead of a DC–DC converter. The output is more filtered and fed as the system power. The battery charging unit controls a rechargeable battery.

Some energy measurement chips have inbuilt power management systems. For an example ADE 5166/ADE5169/ADE5566/ADE5569 chips have inbuilt power management circuitry that manages the regular power supply for battery switchover and power supply failures. Figure 2.17 shows the power supply architecture of the above chips. They are driven by two supply voltages which are V_{DD} and V_{BAT}. V_{DD} is the input voltage from an external supply and V_{BAT} is the battery input. These chips provide automatic battery switchover between V_{DD} and V_{BAT} based on the voltage level detected at V_{DD} or V_{DCIN}. More details on power management system can be obtained referring to the datasheets from Analog Devices website [19].

2.8.4 Energy Measurement Unit

Signal conditioning, ADC, and computation are done inside the energy measurement unit. Energy measurement unit could be a standard energy measurement chip or the system MCU itself. Modern energy measurement chips have digital signal processor (DSP) to perform signal conditioning, ADC and energy calculations. These chips can be found as single phase energy measurement chips or multi phase energy measurement chips. They provide active, reactive, and apparent energy information as data or frequency (pulse) output. RMS voltage measurement, RMS current measurement, frequency, temperature measurement, tampering detection, power management, THD, line SAG detection and communication are also possible in some of them. Some chips operate in single quadrant while others operate in two or four quadrants. They are designed according to IEC and ANSI accuracy standards. Well-known energy measurement chips manufactures are Analog Devices, Microchip, Teridian, STMicroelectronics and NXP.

2.8.5 *Microcontroller*

All functions inside the smart meter are performed by the MCU. It is considered as the core of the meter. It controls the following functions

- Communication with the energy measurement chip
- Calculations based on the data received
- Display electrical parameters, tariff and cost of electricity
- Smartcard reading
- Tamper detection
- Data management with EEPROM
- Communication with other communication devices
- Power management.

Smart meters are normally designed with a LCD. Therefore the consumer is updated with tariff and power outages. Sometimes alarm signals are generated to warn the consumer of higher tariffs and higher demands. Some meters use stepper motor counters rather than a LCD to display the energy consumptions. Those functions are also handled by the external MCU. Some smart meters consist with a single MCU which does all tasks including the energy measurement and routine arithmetic operations. In this case multi-tasking or high degree of parallelism is needed. In other words several operations must be performed simultaneously to the same data sets [1].

2.8.6 *Real Time Clock*

Real time clock is an essential hardware component in all smart meters which keeps track of the current time. It provides time and date information and alarm signals. Some energy measurement chips have a built-in real time clock device. For an example ADE5166 has a built-in real time clock (RTC) which communicates with the internal MCU [19]. Most smart meters have a separately driven RTC unit which can be accessed by the meter MCU. Most of them use reasonable accurate RTCs. The drift has been found as 60 min per year [20]. Smart meters connected to a smart network are periodically synchronized with actual time to avoid time drifts. Meters which are not connected to such network should have a high accurate RTC or should be corrected for the time at regular intervals.

2.8.7 Communicating Systems

The system that consists of smart meters, communication gateway, intelligent control, and data management is known as AMI. Several communication protocols are used in AMI [21]. AMI can consist with a HAN, a Neighborhood Area Network (NAN) and a WAN. Smart meters are the key elements in AMI which need to communicate with domestic appliances, other type of meters (typically water and gas meters), neighboring smart meters, and the energy supplier.

HAN is used to establish a communication link between the smart meter and the smart appliances, other meters, in-home display, and the micro generation unit. HAN provides centralized energy management, services, and facilities. The communication protocol can be a wired or wireless media [1]. Zig-bee, Z-wave, Wi-Fi, and power line communication (PLC) are widely used protocols in HANs. PLC might be a cost effective approach for a HAN but it has many drawbacks due to its robustness. Zig-Bee communication has been recognized as a cost effective, less complexity, low power, and reliable media to handle a HAN [22].

A NAN is used to transfer the data between neighboring smart meters. It facilitates diagnostic messages, firmware upgrades, and real-time messages. Zig-bee communication protocol is widely used in NAN due to high speed of data transferring and low cost [1].

Some smart meters are connected to a remote server through a WAN. They might not be connected to a NAN and data are directly transferred to the server using the wireless media. The communication is established between the meter and the server through a data concentrator for billing purposes, indication of power outages, remote disablement and enablement of supply, security tamper detections, and remote configurations [23]. GSM, GPRS, 3G, and WiMax communication technologies can be used to connect the meter to the WAN. GSM provides wider coverage than other media. However, it will be costly in the long run.

2.9 Smart Meter Standards

There are a lot of standards to measure the accuracy of the metrology of the smart meters. IEC and ANSI standards are commonly used for accuracy measurement of smart meters. Meanwhile standards for other functionalities of the meter like communication are still emerging [24].

Smart meter hardware components and functions should be certified according to the standards. These standards can be internationally or locally developed standards for accuracy, compliance, and functionality criterion. Smart meters are still evolving and many governments, organizations and utility industries are trying to set different standards and guidelines to improve the safety, operability and accuracy of meters and metering devices [11]. For an example the *NEMA SG-AMI* "Requirements for Smart Meter Upgradeability" is being developed by AEIC with NIST and Smart Grid Interoperability Panel (SGIP) in USA. Following are some of standards designed for smart meters in USA.

- Intentional and unintentional radio emissions, and safety related to RF exposure (FCC standards, *parts 1 and 2 of the FCC's Rules and Regulations [47 C.F.R. 1.1307(b), 1.1310, 2.1091, 2.1093.*
- Meter accuracy and performance (*ANSI C12.1, 12.10, and 12.20* specifications)
- Local technical codes and requirements
- Functional tests to satisfy the utilities technical and business requirements
- Utility specifications designed for special area requirements (surge protection for areas vulnerable to lightning, stainless steel enclosures for seaside areas).

Manufactures and utilities do a complete performance test on smart meter hardware and firmware. After the test, the smart meter system components are certified and ready for production and purchase. When it comes to the meter installation several regulations and standards should also be considered. The installation procedure should be done with minimum errors, installation delays and customer issues. Following are some of the regulations for smart meter installation used in USA.

- The National Electric Safety Code (NESC) for utility wiring
- The National Electric Code (NEC) for home wiring
- ASNI C12.1—Code for Electricity Metering
- Local building codes.

After the installation of smart meters, service testing is an essential requirement to maintain the accuracy and performance. Periodic and sample methods are used as in-service testing by utilities. In the periodic method, all the meters are tested on a periodic schedule. Yearly samples of meters are selected in sample method using the manufacturer and purchase date. After the tests, meters should be recalibrated if required [11].

References

1. Ekanayaka J, Liyanage K, Jianzhong W, Yokoyama A, Jenkins N (2012) Smart metering and demand-side integration. In: Smart grid technology and applications, 1st edn. Wiley, Chichester, pp 83–112
2. Harney A (2009) Smart metering technology promotes energy efficiency for a greener world. www.analog.com/library/analogdialogue/…/43…/smart_metering.pdf. Accessed 13 June 2012
3. Bhati NK (2012) Electro-mechanical meters. www.sari-energy.org. Accessed 14 June 2012
4. Electricity meter (2012) en.wikipedia.org/wiki/Electricity_meter. Accessed 14 June 2012

5. Kamakshaiah S, Murthy PK, Amarnath J (2011) Supply meters. In: Electrical measurement and measuring instruments. I. K International Publishing House, India, p 211–295
6. Lession 44 (2013) Study of single phase induction type energy meter or watt hour meter. http://nptel.iitm.ac.in/ Accessed 25 Feb 2013
7. Wan N, Manning K (2001) Exceeding 60-year life expectancy from an electronic energy meter. In: Proceedings metering Asia Pacific conference, 20–22 Feb 2001
8. Wireless system for energy meter reading (2012) www.scribd.com/deeksndeeksd/d/36712865-Abstract-1 Accessed 15 June 2012
9. Jayakrishnan SR (2012) The electrical energy measurement data acquisition engineering based on Scdma. http://seminarprojects.com/Thread-the-electrical-energy-measurment-data-acquisition-engineering-based-on-scdma#ixzz20TRwblMq. Accessed 16 June 2012
10. Ortiz A, Lehtonen M, Manana M, Renedo CJ, Muranen S, Eguiluz LI (2006) Evaluation of energy meters' accuracy based on a power quality test platform. www.sosprocan.unican.es. Accessed 16 June 2012
11. EEI-AEIC-UTC White Paper (2011) Smart meters and smart meter systems: a metering industry perspective. www.aeic.org/meter_service/smartmetersfinal032511.pdf. Accessed 25 Feb 2013
12. ACS712 Data Sheet (2012) www.allegromicro.com/~/Media/…/Datasheets/ACS712-Datasheet.ash. Accessed 17 June 2012
13. Ganesan S (2012) Selection of current transformers and wire sizing in substations. library.abb.com/GLOBAL/SCOT/scot229.NSF/…/gan_pap.pdf. Accessed 18 June 2012
14. Primary Metering and Monitoring current Transformer (2012) http://taehwa.en.ec21.com/Primary_Metering_Monitoring_Current_Transformer–1446971_4342840.html. Accessed 18 June 2012
15. Current shunt Resistors (2012) http://www.rc-electronics-usa.com/current-shunt.html. Accessed 19 June 2012
16. Shepard DE, Yauch DW (2012) An overview of Rogowski coil current sensing technology. www.dynamp.com/dynamp/LDADocum.nsf/…/$FILE/Report.pdf, Accessed 20 June 2012
17. Rogowski Coil (2012) http://www.mfrbee.com/product/685006/Rogowski_Coil.html, Accessed 20 June 2012
18. STMP10 data Sheet (2012) www.st.com/internet/analog/product/250603.jsp. Accessed 21 June 2012
19. ADE 5166 Data Sheet (2012) www.analog.com/en/analog-to-digital…/ade5166/…/product.html. Accessed 21 June 2012
20. MAXIM Smart Meters: Overview (2012) www.maxim-ic.com/solutions/guide/smart-grid/smart-meter.pdf. Accessed 23 June 2012
21. Ahmad S (2011) Smart metering and home automation solutions for the next decade. In: Proceedings of emerging trends in networks and computer communications conference, Udaipur, 22–24 April 2011
22. R. Kopmeiners (2010) Communication Diversity Architecture for Smart Meter Networks, docbox.etsi.org/…smartenergy/kopmeiners_alliander_communication, Accessed 24 June 2012
23. Smart Metering Communications Issues and Technologies (2012) www.cambridgeconsultants.com/downloads/…/smart_metering.pdf. Accessed 24 June 2012
24. The Critical Need for Smart Meter Standards (2013) smartgridsherpa.com/wp-content/…/smart-meters-open-standards.pdf. Accessed 25 Feb 2013

Chapter 3
Basic Functionalities Inside an Energy Measurement Chip

Abstract This chapter consists of functions, calculations, and operations inside the standard energy measurement chips commonly available in the market. Signal conditioning, analog to digital conversion (ADC) and electrical parameter calculations are described step by step where the reader can get an overall idea on digital energy measurement. Signal conditioning includes filtering and effect of aliasing the sampled signal while ADC includes the sampling techniques and quantization noise reduction techniques. The calculation section elaborates the energy measurement, RMS voltage, RMS current, active power, reactive power, apparent power, harmonics measurements, line frequency, power quality measurements, and four quadrant operation. The calculations are based on sinusoidal conditions as well as non-sinusoidal conditions. Different types of energy measurement chips are compared with their energy and power calculations. Digital to frequency conversion (DFC) and communication protocols used in energy measurements chips are also highlighted.

3.1 Introduction

Recent developments in energy measurement have introduced new energy chips with large number of functionalities. Not only the energy measurement but also the RMS voltage, RMS current, active power, reactive power, apparent power, harmonics measurements, line frequency and power quality measurements are possible with modern energy chips. ADCs and DSP are used to perform the signal condition and calculations. They can be found with the International Electrotechnical Commission (IEC) or American National Standards Institute (ANSI) standards. They consume less power and provide high accuracy over wide dynamic range. They get voltage and current signals as input, do all the hard work inside and provide required information through a communication channel. Some chips provide pulse output which is proportional to active and reactive power and

K. S. K. Weranga et al., *Smart Metering Design and Applications*,
SpringerBriefs in Applied Sciences and Technology,
DOI: 10.1007/978-981-4451-82-6_3,

some of them are capable of sending data in binary. However by investigating the architecture of commonly available energy chips in the market, we can identify three main functions inside them. Those functionalities are signal conditioning, ADC and computation.

3.2 Signal Conditioning

The input signals coming from the current and voltage sensors should be amplified and adjusted before ADC. This is called signal conditioning. Basically this stage includes addition/subtraction, attenuation/amplification and filtering [1]. Nowadays energy measurement chips have built-in architecture for this stage. In ADC, proper sampling frequency should be selected to restore the original input waveform. Improper sampling can give rise to aliasing problems which is an artifact of all sampled systems regardless the architecture [2]. Aliasing means that the frequency components greater than half sampling rate shift into the frequency band of desired. This can be reduced by removing the components of the input signal above the Nyquist frequency which is half of the sampling frequency. Therefore it is necessary in all energy measurement design to use an anti aliasing filter to the input signal. This is done by introducing a single pole RC low pass filter.

Figure 3.1 shows the first order analog (RC) low pass filter used in Analog devices ADE5166 energy measurement chip [3].

The gain at a frequency ω is given by (3.1)

$$\frac{V_{out}}{V_{in}} = \frac{1}{1 + j\omega RC}$$
$$\left|\frac{V_{out}}{V_{in}}\right| = \frac{1}{\sqrt{1 + (\omega RC)^2}} \tag{3.1}$$

The transfer function is give by (3.2)

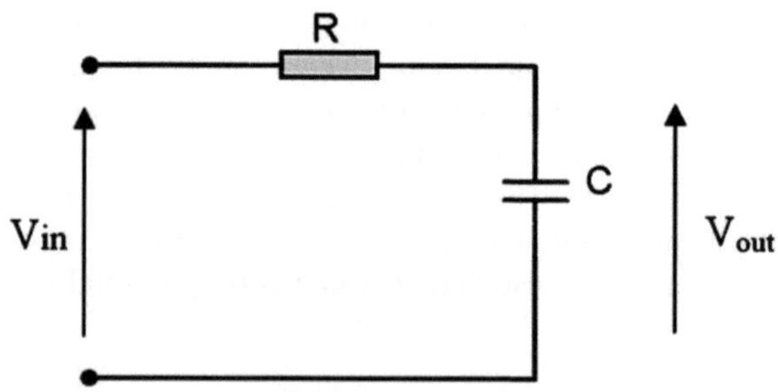

Fig. 3.1 Analog low pass filter

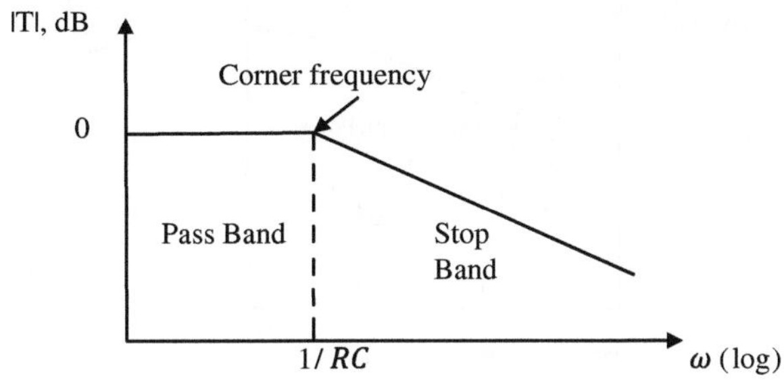

Fig. 3.2 Bode Plot for Eq. (3.2)

$$T(s) = \frac{(1/RC)}{S + (1/RC)} \tag{3.2}$$

The Bode plot for Eq. (3.2) is shown in Fig. 3.2. Significant amount of attenuation can be achieved at higher frequencies using a single pole low pass filter. This RC filter causes a phase shift to the original input signal apart from filtering. This effect can be altered from meter phase calibration.

The sampling frequency may vary from one chip to another. When the chip is designed to measure the fundamental frequency components (voltage, current and power) and harmonic measurements, a sufficient sampling frequency should be selected to obtain the harmonic components accurately [1, 2]. Figure 3.3 illustrates the effect of aliasing discussed in ADE5166 sampling section. The sampling frequency is 819.2 kHz while the Nyquist frequency is 409.6 kHz. Frequency components above the Nyquist frequency are imaged below 409.6 kHz. Frequencies near to 819.2 kHz move into the band of interest of metering. Here the metering band is 40 Hz to 2 kHz. By introducing a simple low pass filter, the high frequencies near to 819.2 kHz can be attenuated [3].

3.3 Analog to Digital Conversion

Analog signals coming from the voltage and current channels are converted to digital at the ADC unit in the energy measurement chip. There are two most popular ADC architectures, successive approximation and sigma-delta method. However sigma-delta method is widely used in modern energy measurement chips. For an example consider the ADE5166 single phase energy measurement chip. It has two sigma-delta ADCs. Sigma-delta converters consist of an interpreter, a single bit digital to analog conversion (DAC), and a latched comparator. The output of these ADCs is used for internal DSP. Figure 3.4 shows the block diagram of a first order sigma-delta modulator and a digital low pass filter [3].

First the output of DAC is subtracted from the input signal coming through the analog low-pass filter. Next the resulting signal is integrated and fed to the comparator. Comparator converts the input signal to a single-bit digital output.

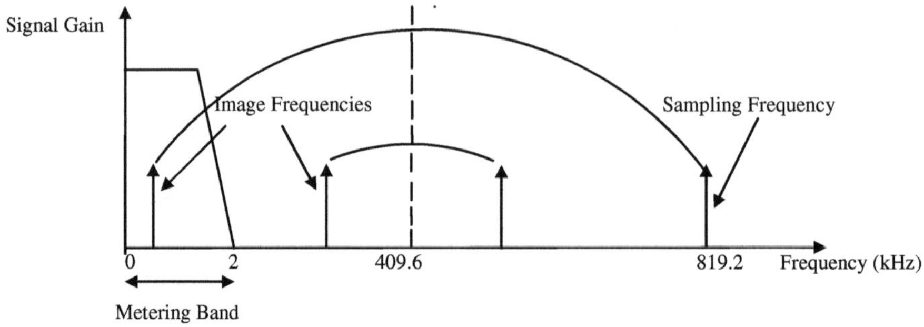

Fig. 3.3 Effect of aliasing (Courtesy of Analog Devices, Inc.)

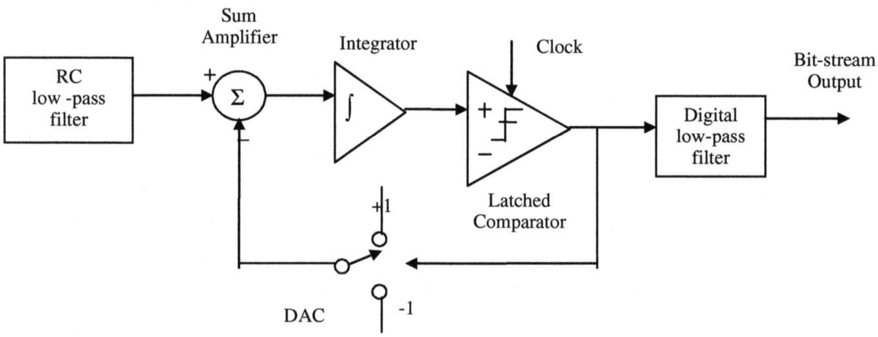

Fig. 3.4 First order sigma-delta modulator (Courtesy of Analog Devices, Inc.)

Finally the latched comparator generates a series of bits corresponding to the analog signal [1, 3].

However ADC causes a quantization noise. This can be explained by modeling the simple sigma-delta modulator as shown in Fig. 3.5. Here the frequency domain analysis is done [4].

The output Y can be written as

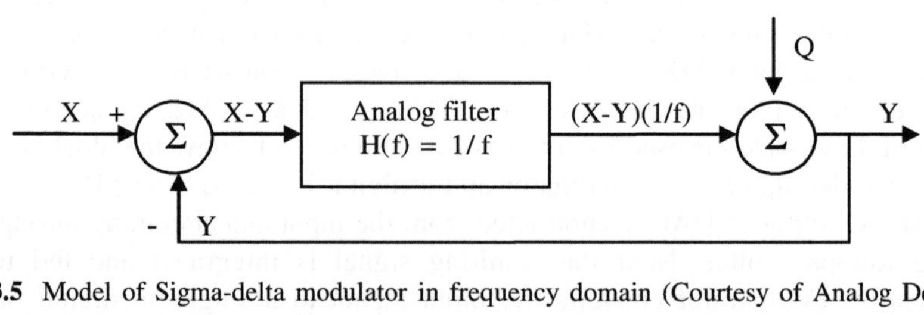

Fig. 3.5 Model of Sigma-delta modulator in frequency domain (Courtesy of Analog Devices, Inc.)

$$Y = \frac{1}{f}(X - Y) + Q \tag{3.3}$$

where:

X is the input and Q is the quantization noise

Rearranging the terms

$$Y = \frac{X}{f+1} + \frac{Qf}{f+1} \tag{3.4}$$

According to (3.4), there are two components.

$\frac{X}{f+1}$ is the signal term and $\frac{Qf}{f+1}$ is the noise term.

Two techniques are used to achieve high resolution in sigma-delta converters. There are

1. oversampling
2. oversampling with shaped quantization noise

In the first case the sampling frequency is selected so that it is many times higher than the bandwidth of interest. For an example ADE5166 has a band of interest 40 Hz to 2 kHz. Therefore the sampling rate is selected as 819.2 kHz which is very much higher than the upper cut off in the band of interest. Furthermore the quantization noise is spread over a wider band due to oversampling. This method lowers the quantization noise in the band of interest [3].

In the second method, the quantization noise is shaped with oversampling. Here the majority of noise lies at higher frequencies. The integrator in the sigma-delta modulator has a high-pass type response for the quantization noise. A digital low pass filter can be added to the output of the sigma-delta modulator to cut off higher frequencies. Figure 3.6 shows the noise reduction due to the oversampling and quantization noise shaping which is used in ADE5166 series energy measurement chips [3].

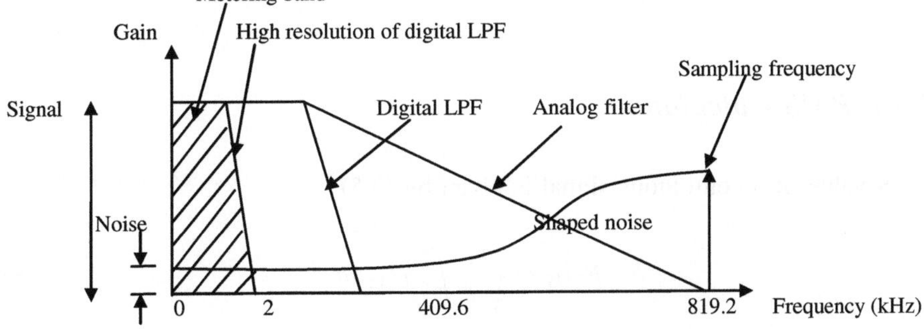

Fig. 3.6 Noise reduction due to oversampling and noise shaping (Courtesy of Analog Devices, Inc.)

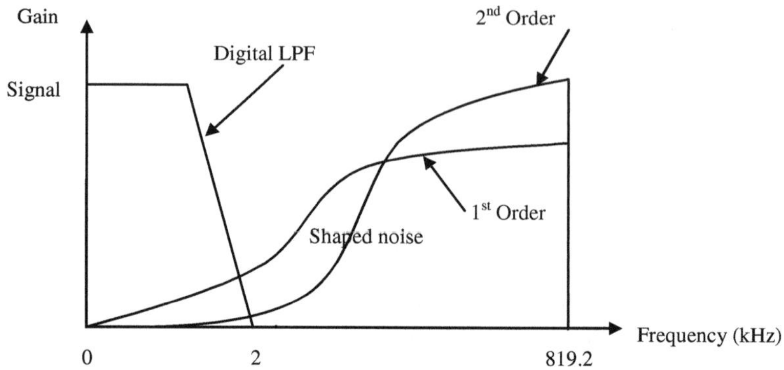

Fig. 3.7 Oversampling with first and second order modulators (Courtesy of Analog Devices, Inc.)

Higher orders of quantization noise shaping can be obtained introducing two or more integration and summing stage for a given oversampling ratio. Figure 3.7 shows the noise shape with the first and the second order modulators [4].

It can be seen that, when the order of the Sigma-Delta modulator is high, the noise that lies in the band of interest is law. Modern energy measurement chips use 2nd order Sigma-Delta ADCs to achieve high accuracy in measurements.

3.4 Basic Electrical Parameters Calculation in a Sinusoidal Single Phase System

The performances and features can vary from energy metering chip to another. Most chips provide general calculation procedure for active power and energy. However some of them are capable of measuring other electrical parameters as well. Following section describes the equations and calculation procedures used in commonly available energy chips in the market.

3.4.1 RMS Calculation

RMS value of a continuous signal is given by (3.5)

$$F_{RMS} = \sqrt{\frac{1}{T}\int_0^T f^2(t)dt} \tag{3.5}$$

where;

F_{RMS} is the RMS value of the function $f(t)$

T is the periodic time

RMS value of a time sampling signal is given by (3.6)

$$F_{rms} = \sqrt{\frac{1}{N}\sum_{n=1}^{N} f_{[n]}^2} \tag{3.6}$$

where;

F_{rms} is the RMS value of the sampling signal

N is the number of samples

n is the nth sample point of the signal

(A) RMS current calculation

To calculate the RMS current of the sampling signal, the following method is used in most energy measurement chips. For an example let's take ADE5166 energy measurement chip. It uses the following method to calculate the RMS current [3].

The instantaneous current signal can be written as

$$I(t) = \sqrt{2} I_{RMS} \sin(\omega t) \tag{3.7}$$

where;

$I(t)$ is referred to the instantaneous current

I_{RMS} is referred to RMS current

ω is referred to angular frequency

t is referred to time

By squaring the both sides and replacing the quantities using trigonometric methods in (3.7)

$$\begin{aligned} I^2(t) &= 2I_{RMS}^2 \sin^2(\omega t) \\ I^2(t) &= I_{RMS}^2 - I_{RMS}^2 \cos(2\omega t) \end{aligned} \tag{3.8}$$

To get the RMS current, the input from the current channel ADC is squared, averaged, and the square root of the average value is taken. The averaging process is done by implementing a low pass filter. According to (3.8), when the signal goes through the low pass filter (LPF), the high frequency component $I_{RMS}^2 \cos(2\omega t)$ is attenuated and the DC term I_{RMS}^2 is extracted [3]. This is illustrated in Fig. 3.8.

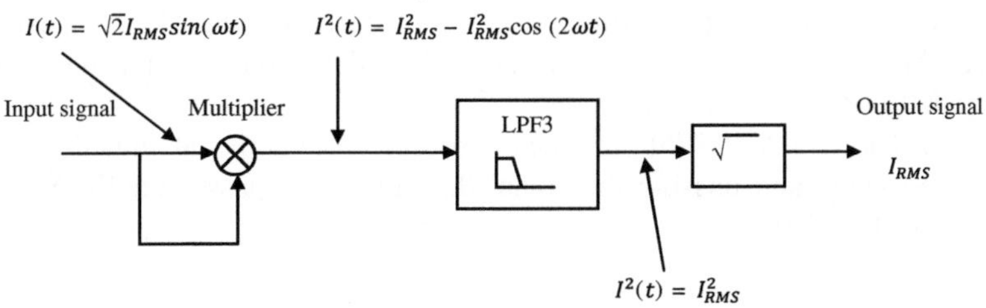

Fig. 3.8 RMS current signal processing in ADE5166 (Courtesy of Analog Devices, Inc.)

(B) RMS voltage calculation
The RMS voltage can be calculated in the same way as the RMS current calculation described in Sect. 3.4.1(A). However some energy measurement chips use mean absolute value calculation to derive the RMS voltage. For an example ADE7758 and ADE5166 use this method [3, 5]. Mean absolute value is accurate when the signal has only the fundamental component. When the signal has harmonics components the calculation is somewhat inaccurate. The mean absolute value calculation can be described as follows.

Let's assume that the voltage signal has only the fundamental component. Therefore the mean absolute value can be written as

$$V_{MAV} = \frac{1}{T}\int_0^T \left|\sqrt{2}V_{RMS}\sin(\omega t)\right| dt \tag{3.9}$$

where;
V_{MAV} is the mean absolute voltage
V_{RMS} is the RMS voltage
T is the periodic time
Considering one cycle of the sine wave we can calculate the V_{MAV}

$$V_{MAV} = \frac{1}{T}\left[\int_0^{T/2} \sqrt{2}V_{RMS}\sin(\omega t)dt - \int_{T/2}^{T} \sqrt{2}V_{RMS}\sin(\omega t)dt\right]$$

Changing the derivatives and the limits to θ form

$$V_{MAV} = \frac{\omega}{2\pi}\left[\int_0^{\pi} \sqrt{2}V_{RMS}\sin\theta\frac{d\theta}{\omega} - \int_{\pi}^{2\pi} \sqrt{2}V_{RMS}\sin\theta\frac{d\theta}{\omega}\right]$$

$$V_{MAV} = \frac{\sqrt{2}V_{RMS}}{2\pi}\left[\int_0^{\pi} \sin\theta d\theta - \int_{\pi}^{2\pi} \sin\theta d\theta\right]$$

$$V_{MAV} = \frac{\sqrt{2}V_{RMS}}{2\pi}\left[-[\cos\theta]_0^{\pi} + [\cos\theta]_{\pi}^{2\pi}\right]$$

$$V_{MAV} = \frac{\sqrt{2}V_{RMS}}{2\pi}[-(-1-1)+(1+1)]$$

$$V_{MAV} = \frac{2\sqrt{2}}{\pi}V_{RMS} \tag{3.10}$$

Thus, V_{MAV} is directly proportional to V_{RMS} . Therefore when V_{MAV} is available, V_{RMS} can be found. However this is valid when the voltage signal has only the fundamental component as mentioned.

In ADE5166 and ADE7758, the mean absolute voltage is calculated by getting the absolute value and passing the signal through a low pass filter [3, 5]. This is illustrated in Fig. 3.9.

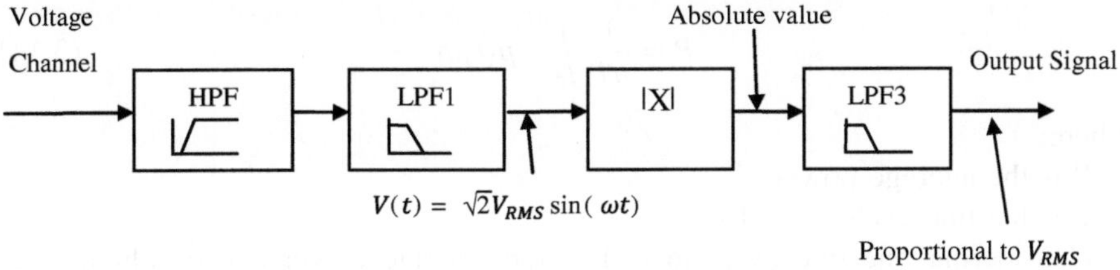

Fig. 3.9 Voltage channel RMS signal processing in ADE5166 (Courtesy of Analog Devices, Inc.)

3.4.2 Active Power and Energy Calculation

(A) Active power calculation

The rate of energy flow from a source to a load is defined as the electrical power. The product of the voltage and the current waveforms gives the instantaneous power signal and it is equal to the rate of energy flow at every instant of time. Several methods are followed to calculate the active power in energy measurement chips.

ADE 5166 and ADE7858 chips use the instantaneous power signal to calculate the active power. The instantaneous power signal is obtained by multiplying the current and voltage signals in each phase. The DC component of the instantaneous power signal is then extracted by LPF2 to obtain the average active power information [3, 5]. This method can be described as follows.

Consider a load with a lagging power factor connected to an AC source. The instantaneous voltage and current can be written as

$$V(t) = \sqrt{2}V_{RMS}\sin(\omega t) \tag{3.11}$$

$$I(t) = \sqrt{2}I_{RMS}\sin(\omega t - \emptyset) \tag{3.12}$$

where;

$V(t)$ and $I(t)$ are instantaneous voltage and current respectively

$\emptyset$ is the phase angle between the voltage and the current

Now the instantaneous power ($p(t)$) can be written as

$$\begin{aligned}
p(t) &= V(t)I(t)\\
p(t) &= \sqrt{2}V_{RMS}\sin(\omega t)\ \sqrt{2}I_{RMS}\sin(\omega t - \emptyset)\\
p(t) &= 2V_{RMS}I_{RMS}\sin(\omega t)\sin(\omega t - \emptyset)\\
p(t) &= V_{RMS}I_{RMS}[\cos(\emptyset) - \cos(2\omega t - \emptyset)]\\
p(t) &= V_{RMS}I_{RMS}\cos(\emptyset) - V_{RMS}I_{RMS}\cos(2\omega t - \emptyset)
\end{aligned} \tag{3.13}$$

The average power over an integral number of line cycles (n) is given by the expression in (3.14).

$$P = \frac{1}{nT}\int_0^{nT} p(t)dt \tag{3.14}$$

where;

P is the average power

T is the line cycle period

Considering one line cycle in (3.14), the average power can be obtained as follows

$$P = \frac{1}{T}\int_0^{T} p(t)dt$$

$$P = \frac{1}{T}\int_0^{T} [V_{RMS}I_{RMS}\cos(\emptyset) - V_{RMS}I_{RMS}\cos(2\omega t - \emptyset)]dt$$

Changing the derivates and the limits to θ form

$$P = \frac{\omega}{2\pi}\int_0^{2\pi} [V_{RMS}I_{RMS}\cos(\emptyset) - V_{RMS}I_{RMS}\cos(2\theta - \emptyset)]\frac{d\theta}{\omega}$$

$$P = \frac{V_{RMS}I_{RMS}}{2\pi}\int_0^{2\pi} [\cos(\emptyset) - \cos(2\theta - \emptyset)]d\theta$$

$$P = \frac{V_{RMS}I_{RMS}}{2\pi}\left[\int_0^{2\pi} \cos(\emptyset)d\theta - \int_0^{2\pi} \cos(2\theta - \emptyset)d\theta\right]$$

$$P = V_{RMS}I_{RMS}\cos(\emptyset) \tag{3.15}$$

Now we can see that active power is equal to the DC component of the instantaneous power signal $p(t)$ from (3.13) and (3.15). Therefore ADE chips use (3.13) to get the active power by extracting the DC component of instantaneous power signal.

However STPM10 energy measurement chips use a different method compared to ADE chips [6]. The functional block diagram of STPM10 is shown in Fig. 3.10.

Consider the instantaneous voltage signal and current signal in (3.11), (3.12). According to the Fig. 3.10,

$$p_1(t) = \frac{dV}{dt}\cdot \int I(t)dt$$

$$p_1(t) = \frac{d\sqrt{2}V_{RMS}\sin(\omega t)}{dt}\cdot \int \sqrt{2}I_{RMS}\sin(\omega t - \emptyset)dt$$

$$p_1(t) = \sqrt{2}V_{RMS}\omega\cos(\omega t)\cdot \sqrt{2}I_{RMS}\left(\frac{-1}{\omega}\right)\cos(\omega t - \emptyset)$$

$$p_1(t) = -2V_{RMS}I_{RMS}\cos(\omega t)\cos(\omega t - \emptyset)$$

$$p_1(t) = -V_{RMS}I_{RMS}[\cos(2\omega t - \emptyset) + \cos(\emptyset)]$$

$$p_1(t) = -V_{RMS}I_{RMS}\cos(\emptyset) - V_{RMS}I_{RMS}\cos(2\omega t - \emptyset) \tag{3.16}$$

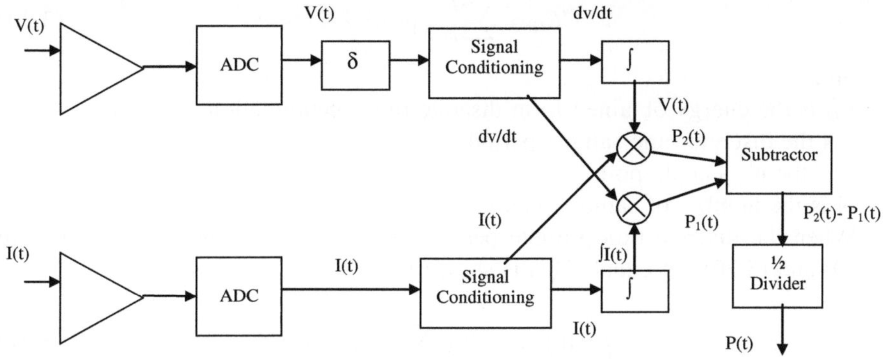

Fig. 3.10 Functional block diagram of STPM10 (Courtesy of STMicroelectronics, Inc.)

According to the Fig. 3.10,

$$\begin{aligned}
p_2(t) &= V(t) \cdot I(t) \\
p_2(t) &= \sqrt{2}V_{RMS}\sin(\omega t) \cdot \sqrt{2}I_{RMS}\sin(\omega t - \emptyset) \\
p_2(t) &= 2V_{RMS}I_{RMS}\sin(\omega t)\sin(\omega t - \emptyset) \\
p_2(t) &= -V_{RMS}I_{RMS}[\cos(2\omega t - \emptyset) - \cos(\emptyset)] \\
p_2(t) &= V_{RMS}I_{RMS}\cos(\emptyset) - V_{RMS}I_{RMS}\cos(2\omega t - \emptyset)
\end{aligned} \tag{3.17}$$

Now active power $p(t)$ can be found by

$$\begin{aligned}
p(t) &= \frac{p_2(t) - p_1(t)}{2} \\
p(t) &= V_{RMS}I_{RMS}\cos(\emptyset)
\end{aligned} \tag{3.18}$$

In this way the active power is calculated in STPM10 energy measurement chip.

(B) Active energy calculation

Active energy is defined as follows

$$E = \int_0^{t_1} p(t)dt \tag{3.19}$$

where;

E is the active energy

$p(t)$ is the instantaneous power

t_1 is the time period of consideration

(3.19) is used for continues time signals. However in energy measurement chips discrete time accumulation is used for active energy calculation. Therefore the active energy is calculated as

$$E_D = \sum\nolimits_{n=0}^{N} \mathrm{p(nt)} \times \mathrm{t} \tag{3.20}$$

where;

E_D is the energy obtained form discrete time accumulation
t is the discrete time sample period
n is the nth sample point
N is the number of sample points

When the discrete time sample period goes to zero, the energy calculated in (3.19) and (3.20) get equal. This is shown in (3.21).

$$\int_0^{t_1} \mathrm{p(t)dt} = \lim_{t \to 0} \sum\nolimits_{n=0}^{N} \mathrm{p(nt)} \times \mathrm{t} \tag{3.21}$$

Equation (3.21) shows that when the discrete time sample period is very small, accuracy of the active energy measurement is high. Therefore energy measurement chip developers try to minimize the error of active energy measurement by keeping the discrete time sample period in a very small value. For an example ADE5166 use 1.22 μs as the discrete time sample period [3]. According to the data sheet of ADE5166, the active energy measurement error is 0.1 % over a 1000:1 dynamic range at 25 °C. MCP3909 energy measurement chip has 1.22 μs of discrete time sample period and its energy measurement error is 0.1 % over a 1000:1 dynamic range [7].

3.4.3 Reactive Power and Energy Calculation

(A) Reactive power calculation

ADE energy measurement chips use the following method to calculate the reactive power. Consider a load with a lagging power factor connected to an AC source. The instantaneous voltage and instantaneous current are denoted in (3.11) and (3.12). For reactive power calculation the instantaneous current signal is shifted 90° and defined as $I'(t)$.

$$I'(t) = \sqrt{2} I_{RMS} \sin\left(\omega t - \emptyset + \frac{\pi}{2}\right) \tag{3.22}$$

Now the instantaneous reactive power signal $(q(t))$ can be obtained by multiplying (3.11) and (3.22)

$$\begin{aligned}
q(t) &= V(t)I'(t)\\
q(t) &= \left[\sqrt{2}V_{RMS}\sin(\omega t)\right]\left[\sqrt{2}I_{RMS}\sin\left(\omega t - \emptyset + \frac{\pi}{2}\right)\right]\\
q(t) &= 2V_{RMS}I_{RMS}\sin(\omega t)\sin\left(\omega t - \emptyset + \frac{\pi}{2}\right)\\
q(t) &= 2V_{RMS}I_{RMS}\sin(\omega t)\cos(\omega t - \emptyset)\\
q(t) &= V_{RMS}I_{RMS}[\sin\emptyset + \sin(2\omega t - \emptyset)]\\
q(t) &= V_{RMS}I_{RMS}\sin\emptyset + V_{RMS}I_{RMS}\sin(2\omega t - \emptyset)
\end{aligned} \tag{3.23}$$

The average reactive power over an integral number of line cycles (n) is given in (3.24).

$$Q = \frac{1}{nT}\int_0^{nT} q(t)dt \tag{3.24}$$

where;

Q is the average reactive power

T is the line cycle period

Considering one line cycle in (3.24), the average reactive power can be obtained as follows

$$Q = \frac{1}{T}\int_0^{T} q(t)dt$$

$$Q = \frac{1}{T}\int_0^{T} [V_{RMS}I_{RMS}\sin\emptyset + V_{RMS}I_{RMS}\sin(2\omega t - \emptyset)]dt$$

By simplification, we get

$$Q = V_{RMS}I_{RMS}\sin\emptyset \tag{3.25}$$

By extracting the DC component of (3.23) we can get the reactive power which is derived in (3.25). Therefore to calculate the reactive power, the voltage signal and 90° phase-shifted current signal are multiplied and low pass filtered [3, 5].

The STPM10 chips follow a different method to calculate the reactive power compared to ADE chips [6]. This method is shown in the block diagram of Fig. 3.10. Consider the instantaneous voltage and current signal in (3.11), (3.12).

$$\begin{aligned}
Q_1(t) &= V(t)\cdot\int I(t)dt\\
Q_1(t) &= \sqrt{2}V_{RMS}\sin(\omega t)\cdot\int\sqrt{2}I_{RMS}\sin(\omega t - \emptyset)dt\\
Q_1(t) &= \sqrt{2}V_{RMS}\sin(\omega t)\cdot\sqrt{2}I_{RMS}\left(\frac{-1}{\omega}\right)\cos(\omega t - \emptyset)
\end{aligned}$$

$$Q_1(t) = 2\left(\frac{-1}{\omega}\right) V_{RMS} I_{RMS} \sin(\omega t) \cdot \cos(\omega t - \emptyset)$$

$$Q_1(t) = \frac{1}{\omega} V_{RMS} I_{RMS} [-\sin(2\omega t - \emptyset) - \sin(\emptyset)] \tag{3.26}$$

Multiplying RHS by ω in (3.26), we can get

$$Q_2(t) = V_{RMS} I_{RMS} [-\sin(2\omega t - \emptyset) - \sin(\emptyset)]$$

$$Q_2(t) = -V_{RMS} I_{RMS} \sin(\emptyset) - V_{RMS} I_{RMS} \sin(2\omega t - \emptyset) \tag{3.27}$$

The DC component of (3.27) can be used to get the reactive power. The negative value in the reactive power is due to the lagging power factor which was originally considered for the derivation. Therefore the sign should be changed according to the type of the load. Moreover STPM10 chip requires a separate calculation for ω in order to find the reactive power.

(B) Reactive energy calculation

Reactive energy is defined as follows

$$E_R = \int_0^{t_1} q(t)dt \tag{3.28}$$

where;

E_R is the reactive energy

$q(t)$ is the instantaneous reactive power

t_1 is the time period of consideration

Energy measurement chips use discrete time accumulation for reactive energy calculation. Therefore reactive energy can be written for discrete time accumulation mode as in (3.29)

$$E_R = \lim_{t \to 0} \sum_{n=0}^{N} q(nt) \times t \tag{3.29}$$

where;

t is the discrete time sample period

n is the nth sample point

N is the number of sample points

Practically the discrete time sample period cannot be kept as zero. Therefore energy measurement chip developers try to minimize the error of reactive energy measurement by keeping the discrete time sample period in a very small value. For an example ADE5166 uses 1.22 μs as the discrete time sample period for reactive energy calculation [3].

3.4.4 Apparent Power and Energy Calculation

(A) Apparent power calculation
ADE7758 chip uses an arithmetic approach for the apparent power calculation [5]. Therefore to get the apparent power, the RMS voltage and the RMS current are multiplied as shown in (3.30).

$$S = V_{RMS} I_{RMS} \tag{3.30}$$

where;

S is the arithmetic apparent power

The output from the multiplier is then low-pass filtered to get the average apparent power.

(B) Apparent energy calculation
Apparent energy (E_A) can be obtained according to (3.31)

$$E_A = \int_0^{t_1} S(t)dt \tag{3.31}$$

where;

E_A is the apparent energy

$S(t)$ is the instantaneous apparent power

t_1 is the time period of consideration

The same procedure is used to calculate the apparent energy as described in ADE7758 active and reactive energy calculation [5]. Discrete time accumulation is used as shown in (3.32) in this chip.

$$E_A = \lim_{t \to 0} \sum\nolimits_{n=0}^{N} S(nt) \times t \tag{3.32}$$

where;

t is the discrete time sample period

n is the nth sample point

N is the number of sample points

3.5 Power Quality Measurements

Power quality is a measurement of fitness of the electrical power delivered to consumer loads. Power quality issues are considerable facts in industries and commercial sectors. Low quality power results failure or malfunction of equipments resulting problems in productions. The most common power quality measurements are voltage variations (long durations), voltage fluctuations, voltage unbalance, voltage sag, transient over voltages, and harmonics [8]. Therefore measurement of these quantities is an essential part in order to maintain a high quality power. Modern energy measurement chips are capable of measuring above quantities. Most of the Analog devices energy measurement chips (ADE series chips) have a separate mechanism to monitor the quality of the power. These chips perform frequency calculation, line voltage sag detection, peak voltage detection, and peak current detection. For the frequency analysis, zero crossing detection or Fourier analysis is used. Three phase energy measurement chips are capable of phase sequence detection. This is done by zero crossing detection and comparison [3, 5]. However ADE7880 energy measurement chip is capable of measuring fundamentals and harmonics of a three phase system. This chip uses second order Sigma-Delta ADCs. Total harmonic distortion is also computed for current and voltage in each phase [9].

3.6 Electrical Parameters Calculation in a Non-sinusoidal Single Phase System

In the real word, many systems consist of non linear loads which cause harmonics in the supply system. Many researchers and organizations have put a lot of effort to clarify the definitions of power quantities in non-sinusoidal systems. The IEEE Std 1459 was published in 2010 by power system instrumentation and measurement committee of IEEE to provide information on measuring electrical parameters under sinusoidal, non-sinusoidal, balanced or unbalanced load systems. Following section describes the electrical quantity measurement in a single phase system under non-sinusoidal waveforms according to the IEEE Std 1459 which is very useful for electrical instrumentation [10].

In steady state conditions periodic instantaneous voltage and current have power system frequency components, DC components and harmonic components. Therefore the instantaneous voltage and current can be written as

$$v(t) = V_0 + \sqrt{2}V_1 \sin(\omega t - \alpha_1) + \sqrt{2}\sum\nolimits_{h=2}^{\infty} V_h \sin(h\omega t - \alpha_h) \tag{3.33}$$

$$i(t) = I_0 + \sqrt{2}I_1 \sin(\omega t - \beta_1) + \sqrt{2}\sum\nolimits_{h=2}^{\infty} I_h \sin(h\omega t - \beta_h) \tag{3.34}$$

where;

h is the harmonic order
V_1 is the RMS of fundamental voltage
V_0 is the DC voltage
V_h is the RMS of the hth harmonic voltage
I_1 is the RMS of fundamental current
I_0 is the DC current
I_h is the RMS of the hth harmonic current
The RMS value of the voltage is given by

$$V_{RMS} = \sqrt{\frac{1}{T}\int_0^T v^2(t)dt} = \sqrt{V_1^2 + V_0^2 + \sum\nolimits_{h=2}^{\infty} V_h^2} = \sqrt{V_1^2 + V_H^2} \tag{3.35}$$

where;

$$V_H^2 = V_0^2 + \sum\nolimits_{h=2}^{\infty} V_h^2 \tag{3.36}$$

The RMS value of the current is given by

$$I_{RMS} = \sqrt{\frac{1}{T}\int_0^T i^2(t)dt} = \sqrt{I_1^2 + I_0^2 + \sum\nolimits_{h=2}^{\infty} I_h^2} = \sqrt{I_1^2 + I_H^2} \tag{3.37}$$

where;

$$I_H^2 = I_0^2 + \sum\nolimits_{h=2}^{\infty} I_h^2 \tag{3.38}$$

The quality of the power system is determined by the power quality measurements. They are basically the total harmonic distortion of voltage and current. These values help to determine the overall deviation of a distorted wave from its fundamental.

Total harmonic distortion of the voltage is given by

$$THD_V = \sqrt{\left(\frac{V}{V_1}\right)^2 - 1} \tag{3.39}$$

where;

THD_V is the total harmonic distortion of the voltage and V is the RMS of supply voltage

Total harmonic distortion of the current is

$$THD_I = \sqrt{\left(\frac{I}{I_1}\right)^2 - 1} \tag{3.40}$$

where;

THD_I is the total harmonic distortion of the current and I is the RMS of supply current

The instantaneous power can be obtained from the product of voltage and current signal

$$p(t) = v(t)i(t)$$

$$\begin{aligned} p(t) = &\left[V_0 + \sqrt{2}V_1 \sin(\omega t - \alpha_1) + \sqrt{2}\sum\nolimits_{h=2}^{\infty} V_h \sin(h\omega t - \alpha_h)\right] \\ &\times \left[I_0 + \sqrt{2}I_1 \sin(\omega t - \beta_1) + \sqrt{2}\sum\nolimits_{h=2}^{\infty} I_h \sin(h\omega t - \beta_h)\right] \end{aligned} \tag{3.41}$$

By simplification we can get

$$
\begin{aligned}
p(t) &= V_0 I_0 + \sum_{h=1}^{\infty} V_h I_h \cos(\beta_h - \alpha_h)[1 - \cos(2h\omega t - 2\alpha_h)] \\
&- \sum_{h=1}^{\infty} V_h I_h \sin(\beta_h - \alpha_h) \sin 2(h\omega t - \alpha_h) \\
&+ 2\sum_{n=1}^{\infty} \sum_{\substack{m=1 \\ m \neq n}}^{\infty} V_m I_n \sin(m\omega t - \alpha_m) \sin(n\omega t - \beta_n) \\
&+ \sqrt{2} V_0 \sum_{h=1}^{\infty} I_h \sin(h\omega t - \beta_h) + \sqrt{2} I_0 \sum_{h=1}^{\infty} V_h \sin(h\omega t - \alpha_h)
\end{aligned}
\tag{3.42}
$$

where;

$p(t)$ is the instantaneous power signal under non-sinusoidal wave forms

In other words

$$p(t) = p_a + p_q \tag{3.43}$$

where;

$$
\begin{aligned}
p_a &= V_0 I_0 + \sum\nolimits_{h=1}^{\infty} V_h I_h \cos(\beta_h - \alpha_h)[1 - \cos(2h\omega t - 2\alpha_h)] \\
&= V_0 I_0 + \sum\nolimits_{h=1}^{\infty} V_h I_h \cos(\beta_h - \alpha_h) - \sum\nolimits_{h=1}^{\infty} V_h I_h \cos(2h\omega t - 2\alpha_h)
\end{aligned}
\tag{3.44}
$$

p_a is the part of the instantaneous power that is equal to the sum of harmonic active powers. The terms $V_0 I_0$ and $V_h I_h \cos(\beta_h - \alpha_h)$ contribute to net transfer of energy while $V_h I_h \cos(2h\omega t - 2\alpha_h)$ doesn't contribute to net transfer of energy.

$$p_q = -\sum\nolimits_{h=1}^{\infty} V_h I_h \sin(\beta_h - \alpha_h) \sin 2(h\omega t - \alpha_h) \tag{3.45}$$

p_q is a term that does not represent a net transfer of energy

The active power can be obtained by integrating the instantaneous power signal and getting the average value over number of line cycles. By simplification we can obtain that the active power is equal to the sum of fundamental active power and harmonic active power. Therefore we can write

$$P = P_1 + P_H \tag{3.46}$$

where;

P is the active power

P_1 is the fundamental active power

P_H is the harmonic active power or the nonfundamantal active power

And

$$P_1 = \frac{1}{nT}\int_0^{nT} v_1(t)i_1(t)\,dt = V_1 I_1 \cos(\beta_1 - \alpha_1) \tag{3.47}$$

where;

$v_1(t)$is the fundamental instantaneous voltage

$i_1(t)$is the fundamental instantaneous current

$$P_H = V_0 I_0 + \sum\nolimits_{h=2}^{\infty} V_h I_h \cos(\beta_h - \alpha_h) \tag{3.48}$$

Fundamental reactive power is given by

$$Q_1 = \frac{\omega}{nT}\int_0^{nT} i_1 \left[\int v_1 dt\right] dt = V_1 I_1 \sin(\beta_1 - \alpha_1) \tag{3.49}$$

where;

Q_1 is the fundamental reactive power

According to (3.30), the apparent power is the product of RMS voltage and RMS current. The apparent power also can be described in two terms which are fundamental apparent power and non-fundamental apparent power.

Fundamental apparent power can be obtained from the product of fundamental voltage and fundamental current. This is also equals to the square root of squares of fundamental active and fundamental reactive power.

$$S_1 = V_1 I_1 = \sqrt{P_1^2 + Q_1^2} \tag{3.50}$$

where;

S_1 is the fundamental apparent power

Non-fundamental apparent power can be written as

$$S_N = \sqrt{S^2 - S_1^2} \tag{3.51}$$

where;

S_N is the non-fundamental apparent power and S is the arithmetic apparent power.

There is also a term called as non-active power. This power lumps together both fundamental and non-fundamental non-active components. It was called as "fictitious power" in the past. Non-active power is equal to the reactive power when the voltage and current waveforms are exactly sinusoidal. The non-active power can be defined as

$$N = \sqrt{S^2 - P^2} \tag{3.52}$$

where;

N is the non-active power

There are other types of power derived under harmonics. They are current distortion power, voltage distortion power, and harmonic apparent power.

Current distortion power is given by

$$D_I = V_1 I_H = S_1(THD_I) \tag{3.53}$$

where;
D_I is the current distortion power
Voltage distortion power

$$D_V = V_H I_1 = S_1(THD_V) \tag{3.54}$$

where;
D_V is the voltage distortion power
Harmonic apparent power

$$S_H = V_H I_H = S_1(THD_I)(THD_V) \tag{3.55}$$

where;
S_H is the harmonic apparent power
Power factor is also a term which determines how usefully the power is consumed. In non-sinusoidal systems we have two terms which are fundamental power factor and power factor.

The fundamental power factor is helpful to analyze the fundamental power flow conditions. It is the cosine of the phase angle between the fundamental voltage and the fundamental current. It is also equals to the ratio between the fundamental active and fundamental apparent power. The fundamental power factor can be written as

$$PF_1 = \cos(\beta_1 - \alpha_1) = \frac{P_1}{S_1} \tag{3.56}$$

where;
PF_1 is the fundamental power factor
The Power factor for a harmonic system is given by

$$PF = \frac{P}{S} = \frac{P_1 + P_H}{\sqrt{S_1^2 + S_N^2}} \tag{3.57}$$

where;
PF is the power factor
According to the IEEE 1459™-2000 electric parameters in a non sinusoidal single phase system can be summarized as in the Table 3.1 [10].

Table 3.1 IEEE 1459™-2000 electric parameters in a non sinusoidal single phase system

Electric quantity	SI unit	Fundamental	Non-fundamental	Combined
Active power	W	P_1	P_H	P
Non-active power	var	Q_1	D_1, D_V, D_H	Q
Apparent power	VA	S_1	S_N, S_H	S
Line utilization	–	$PF_1 = P_1/S_1$	–	PF = P/S
Harmonic pollution	–	–	–	S_N/S_1

3.6.1 Harmonics and Energy Measurement in ADE 7880 Energy Measurement Chip

ADE 7880 is a high accurate 3 phase energy measurement chip which supports IEC 62053-21, IEC 62053-22, IEC 62053-23, EN 50470-1, EN 50470-3, ANSI C12.20, and IEEE1459 standards. This chip has been used in energy metering systems, power quality monitoring, solar inverters, process monitoring, and protective devices due to its complete harmonic analysis. It measures RMS voltage, RMS current, all harmonics (up to the 63rd harmonic), active, reactive and apparent powers, power factor, THD, harmonic distortion, total (fundamental and harmonic) active and apparent energy and fundamental active/reactive energy on each phase [9]. In this section electrical parameter calculation inside ADE 7880 energy measurement chip is discussed.

(A) RMS voltage and current calculation under non-sinusoidal waveforms
Let us take the time varying voltage or current signal as $f(t)$which contains power system frequency component, DC component and harmonic components.

$$f(t) = F_0 + \sqrt{2}F_1 \sin(\omega t - \alpha_1) + \sqrt{2}\sum\nolimits_{h=2}^{\infty} F_h \sin(h\omega t - \alpha_h) \tag{3.58}$$

Or

$$f(t) = F_0 + \sqrt{2}\sum\nolimits_{h=1}^{\infty} F_h \sin(h\omega t - \alpha_h) \tag{3.59}$$

where;

$f(t)$ is the time varying voltage or current signal
F_1 is the RMS of fundamental signal
F_0 is the DC signal
F_h is the RMS of the hth harmonic signal
In the chip, this f(t) signal is squared, low pass filtered and taken the square root of the results. The procedure can be explained as follows.
First the voltage or the current signal is squared.

$$f^2(t) = \left[F_0 + \sqrt{2}\sum\nolimits_{h=1}^{\infty} F_h \sin(h\omega t - \alpha_h)\right]\left[F_0 + \sqrt{2}\sum\nolimits_{h=1}^{\infty} F_h \sin(h\omega t - \alpha_h)\right] \tag{3.60}$$

$$f^2(t) = F_0^2 + 2\sqrt{2}F_0\sum_{h=1}^{\infty} F_h\sin(h\omega t - \alpha_h) + \left[\sqrt{2}\sum_{h=1}^{\infty} F_h\sin(h\omega t - \alpha_h)\right]^2$$

$$f^2(t) = F_0^2 + 2\sqrt{2}F_0 \sum_{h=1}^{\infty} F_h \sin(h\omega t - \alpha_h) + 2\sum_{h=1}^{\infty} F_h^2 \sin^2(h\omega t - \alpha_h)$$
$$+ 2 \sum_{\substack{h, m = 1 \\ h \neq m}}^{\infty} 2F_hF_m \sin(h\omega t - \alpha_h)\sin(m\omega t - \alpha_h)$$

$$f^2(t) = F_0^2 + 2\sqrt{2}F_0 \sum_{h=1}^{\infty} F_h \sin(h\omega t - \alpha_h) + \sum_{h=1}^{\infty} F_h^2[1 - \cos 2(h\omega t - \alpha_h)]$$
$$+ 2 \sum_{\substack{h, m = 1 \\ h \neq m}}^{\infty} 2F_hF_m \sin(h\omega t - \alpha_h)\sin(m\omega t - \alpha_h)$$

$$f^2(t) = F_0^2 + \sum_{h=1}^{\infty} F_h^2 + 2\sqrt{2}F_0 \sum_{h=1}^{\infty} F_h \sin(h\omega t - \alpha_h) - \sum_{h=1}^{\infty} F_h^2 \cos 2(h\omega t - \alpha_h)$$
$$+ 2 \sum_{\substack{h, m = 1 \\ h \neq m}}^{\infty} 2F_hF_m \sin(h\omega t - \alpha_h)\sin(m\omega t - \alpha_h) \tag{3.61}$$

This signal is passed through a low pass filter. Then the output will be

$$F_{RMS}^2 = F_0^2 + \sum\nolimits_{h=1}^{\infty} F_h^2 \tag{3.62}$$

All the time varying terms (ω *terms*) are get cancelled after low pass filtering. RMS value of the signal is obtained by taking the square root of (3.62)

$$F_{RMS} = \sqrt{F_0^2 + \sum\nolimits_{h=1}^{\infty} F_h^2} \tag{3.63}$$

(B) Total active power and energy calculation under non-sinusoidal waveforms
According to (3.41), the instantaneous power in a single phase AC system is the product of voltage and current signals. To get the active power or the real power, this instantaneous power signal should be integrated over line cycles and then averaged. Therefore the average power can be found as

$$P = \frac{1}{nT}\int_0^{nT} p(t)dt = V_0I_0 + \sum\nolimits_{h=1}^{\infty} V_hI_h \cos(\beta_h - \alpha_h) \tag{3.64}$$

According to (3.64) and (3.44), we can see that the active power is equal to the DC component of the instantaneous power signal.

The ADE 7880 energy measurement chip uses this method to calculate the active power in each phase. First, the instantaneous power signal is obtained by multiplying the current and voltage signal in each phase. Then this signal is low pass filtered to obtain the DC component which is equal to the active power in each phase. This chip has separate mechanism to isolate the active powers in each harmonic. The user can select the order of harmonics to be monitored and can read the power in required harmonic by accessing to the internal readable registers.

The active energy in each phase is calculated by using (3.65).

$$e = \int_0^{nT} p(t)dt = nT\left[V_0 I_0 + \sum\nolimits_{h=1}^{\infty} V_h I_h \cos(\beta_h - \alpha_h)\right] \tag{3.65}$$

where;

e is the active energy

(C) Fundamental reactive power and energy calculation under non-sinusoidal waveforms

ADE 7880 energy measurement chip has the capability to measure the fundamental reactive power as well as the harmonic reactive powers in each phase. First, the instantaneous reactive power signal is generated. It is done by multiplying the each harmonics of the voltage signals by phase shifted (90° phase shift) corresponding harmonic currents. Then this signal is low pass filtered to extract the DC component which is equal to the summation of each harmonic reactive power. A proprietary algorithm is used to compute the fundamental reactive power in the chip. This can be explained as follows.

The 90° phase shifted current can be written as

$$i'(t) = I_0 + \sum\nolimits_{h=1}^{\infty} \sqrt{2} I_h \sin\left(h\omega t - \beta_h + \frac{\pi}{2}\right) \tag{3.66}$$

where;

$i'(t)$ is the 90° phase shifted current signal

Then the instantaneous reactive power signal is

$$q(t) = v(t)i'(t)$$

$$q(t) = \left[V_0 + \sum_{h=1}^{\infty} \sqrt{2} V_h \sin(h\omega t - \alpha_h)\right]\left[I_0 + \sum_{h=1}^{\infty} \sqrt{2} I_h \sin(h\omega t - \beta_h + \frac{\pi}{2})\right]$$

$$q(t) = V_0 I_0 + V_0 \sum_{h=1}^{\infty} \sqrt{2} I_h \sin\left(h\omega t - \beta_h + \frac{\pi}{2}\right) + I_0 \sum_{h=1}^{\infty} \sqrt{2} V_h \sin(h\omega t - \alpha_h)$$
$$+ \sum_{h=1}^{\infty} V_h I_h 2 \sin(h\omega t - \alpha_h) \sin\left(h\omega t - \beta_h + \frac{\pi}{2}\right)$$
$$+ \sum_{\substack{h,m=1 \\ n \neq m}}^{\infty} V_h I_m 2 \sin(h\omega t - \alpha_h) \sin\left(m\omega t - \beta_h + \frac{\pi}{2}\right)$$

$$q(t) = V_0 I_0 + V_0 \sum_{h=1}^{\infty} \sqrt{2} I_h \sin\left(h\omega t - \beta_h + \frac{\pi}{2}\right)$$
$$+ I_0 \sum_{h=1}^{\infty} \sqrt{2} V_h \sin(h\omega t - \alpha_h)$$
$$+ \sum_{h=1}^{\infty} V_h I_h \left[\cos\left(\beta_h - \alpha_h - \frac{\pi}{2}\right) - \cos\left(2h\omega t - \alpha_h - \beta_h + \frac{\pi}{2}\right)\right]$$
$$+ \sum^{\infty} {}_{\substack{n,m = 1 \\ n \neq m}} V_n I_m \left[\cos\left((h-m)\omega t - \alpha_h + \beta_h - \frac{\pi}{2}\right)\right.$$
$$\left. - \cos\left((h+m)\omega t - \alpha_h - \beta_h + \frac{\pi}{2}\right)\right] \tag{3.67}$$

where;

$q(t)$ is the instantaneous reactive power signal under non-sinusoidal waveforms

The average total reactive power can be obtained according to (3.68)

$$Q = \frac{1}{nT} \int_0^{nT} q(t)dt = V_0 I_0 + \sum_{h=1}^{\infty} V_h I_h \cos\left(\beta_h - \alpha_h - \frac{\pi}{2}\right)$$
$$= V_0 I_0 + \sum_{h=1}^{\infty} V_h I_h \sin(\beta_h - \alpha_h) \tag{3.68}$$

After the low pass filtering the frequency terms of (3.67) are filtered out and the DC component is extracted. This gives the total reactive power in one phase.

3.7 Four Quadrant Operation

Some energy chips measure both positive and negative active energy. Some of them are also capable of measuring both positive and negative active and reactive energy which is known as four quadrant operation. For an example ADE 7953, ADE 7758, MCP 3909, and 78M6613 are few energy measurement chips which support four quadrant operation. They have different registers to determine the polarity of power. By accessing these registers the direction of power flow can be determined. Using an external MCU, the active power inflow, active power out

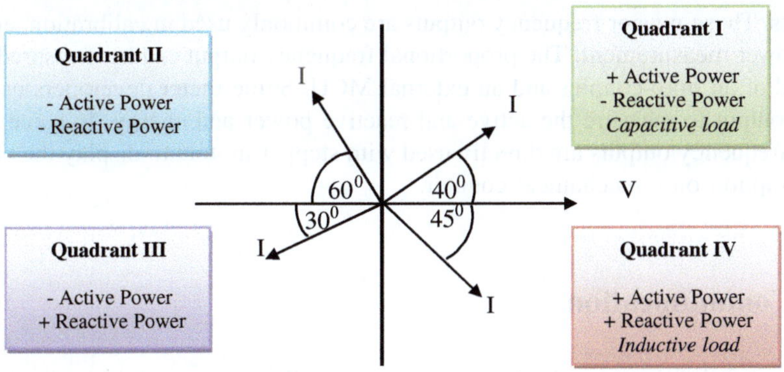

Fig. 3.11 Four quadrant operation

flow, reactive power inflow and reactive power outflow can be separately measured, stored and displayed.

Figure 3.11 illustrates the sign of active and reactive power with the voltage and current directions.

As shown in the Fig 3.11, the active power measurement is positive when the phase angle between the voltage and the current is between 0° and 90° (quadrant I), and between 270° and 360° (quadrant IV). In quadrant I and quadrant IV the active energy is delivered from the source to the consumer. The active power is negative when the phase angle between the voltage and the current is between 90° and 180° (quadrant II), and between 180° and 270° (quadrant III). In quadrant II and quadrant III, the active energy is delivered back to the system from the consumer.

The reactive power measurement is positive when the phase angle between the voltage and the current is between 180° and 270° (quadrant III), and between 270° and 360° (quadrant IV). The reactive power is delivered to the consumer in quadrant III and quadrant IV. When the phase angle between the voltage and current is between 0° and 90° (quadrant I), and between 90° and 180° (quadrant II), the reactive power measured is negative. In quadrant I and II reactive power is received to the source from the consumer.

The connected load of the consumer is capacitive in quadrant I and the power factor is considered as leading. The power factor is lagging when the load is inductive in quadrant IV.

3.8 Frequency Outputs

Modern energy measurement chips provide frequency output for active, reactive and apparent power. They use a digital to frequency converter (DFC) to generate a frequency proportional to the power under steady state load conditions. Specific pins are allocated for active power, reactive power, and apparent power frequency outputs. These pins or frequency outputs are commonly used in calibration, energy and power measurement. The proportional frequency output can be measured with the aid of an opto-coupler and an external MCU. Some meter developers use this pulse output to measure the active and reactive power and energy. In some cases these frequency outputs are directly used with stepper motors to display the energy consumption on a mechanical counter.

3.9 Communication

Communication is a mandatory requirement in all energy measurement systems. The data processed inside the energy chips including voltage, current, power, energy, frequency, and power quality measurements should be transmitted to an external MCU. A communication link is established between the energy measurement chip and the external MCU for data handling and calibration. Serial peripheral interface (SPI) and RS232 are the widely used communication protocols in energy measurement chips. The internal registry data which contain the electrical parameters are accessed by external MCU through these links. For an example Analog devices, microchip, and STMicroelectronics provide energy measurement chips with a built-in SPI interface while Maxim chips provide a UART interface.

References

1. Ekanayaka J, Liyanage K, Jianzhong W, Yokoyama A, Jenkins N (2012) Smart metering and demand-side integration. In: Smart grid technology and applications, 1st edn, Wiley, Chichester, pp 83–112
2. Kumarawadu S (2010) DSPs in control systems. In: Control systems theory and implementations, 1st edn, Narosa Publishing House, India, pp 122–147
3. ADE 5166 Data Sheet, www.analog.com/en/analog-to-digital…/ade5166/…/product.html. Accessed 21 June 2012
4. Kester W, Bryant J, Buxton J, ADCs for signal conditioning, www.analog.com/static/imported-files/…/732529616sscsect8.PDF. Accessed 02 July 2012
5. ADE 7758 Data Sheet, www.analog.com/static/imported-files/data_sheets/ADE7758.pdf. Accessed 03 July 2012
6. STMP10 data Sheet, www.st.com/internet/analog/product/250603.jsp. Accessed 21 June 2012
7. MCP 3909 Data Sheet, ww1.microchip.com/downloads/en/DeviceDoc/22025a.pdf. Accessed 03 July 2012
8. Power Quality, Available: http://en.wikipedia.org/wiki/Power_quality. Accessed 24 Feb 2013
9. ADE 7880 Data Sheet, www.analog.com/static/imported-files/data_sheets/ADE7880.pdf. Accessed 03 July 2012
10. IEEE Std 1459™-2010 IEEE Power and Energy Society. www.ieee.org. Accessed 17 Feb 2013

Chapter 4
Smart Meter Prototype Design

Abstract An example of smart meter prototype development is discussed with software and hardware architectures. By selecting a standard energy measurement chip (ADE7758) together with a microcontroller (PIC 18F452), a real time clock IC (PCF8583), and a GSM module (SIM900), the prototype is designed. Hardware configurations and software algorithms are illustrated and discussed so that the reader can easily read and understand the real structure of a smart meter. Three phase, per phase, time and date information are displayed on a LCD at the meter side. The data processed at the smart meter are sent to a remote server in every 15 min intervals. Consumption details including the electricity cost, average energy usage, average daily cost, predicted electricity cost of the month, and the number of remaining days for specified credit levels are calculated and displayed at the server side. This smart metering system illustrates the two way interaction between the consumer and the utility together with information service.

4.1 Introduction

In this chapter we have described how to design a smart meter (a prototype) using a standard energy measurement chip. This design basically consists of an energy measurement chip, microcontroller, real time clock, LCD and a GSM module. To measure the power, voltage and other electrical parameters the energy measurement chip ADE 7758 is used. This chip is highly accurate (0.1 % error in active energy) over the selected current and voltage range and it supports the IEC 60687 and other IEC standards. This chip is capable of measuring electrical parameters in a poly-phase system. Three current transformers are used as current sensors while three resistor dividers are used as voltage sensors. The reference voltage is kept at 2.4 V. Here we have followed the test circuit of ADE7758 which is available in the data sheet and the schematic is shown in Appendix 1 [1].

K. S. K. Weranga et al., *Smart Metering Design and Applications*,
SpringerBriefs in Applied Sciences and Technology,
DOI: 10.1007/978-981-4451-82-6_4,

The energy chip is connected to a PIC microcontroller (18F452) for communication and calibration. The microcontroller receives the data via the SPI bus from the energy chip and displays at the demand side using a LCD. The user has the freedom to monitor phase parameters as well as energy usage with a push button located at the bottom of the meter. The user can lock/unlock data in a particular phase using the other push button. The reset button is used to reset the energy information in the EEPROM.

The real time clock module is designed with PCF8583 IC. It provides timely data and alarm signals to the MCU via I2C bus. SIM 900 GSM module is used for the data transmission. It communicates with the MCU via RS 232 channel. The power supply is designed to provide adequate power to all components of the system. Even though this prototype is specially designed for domestic three phase applications, it can be modified even for an industrial. application. Figure 4.1 shows the block diagram of the smart meter design.

4.2 Basic Operation of ADE7758 Energy Chip

All the analog inputs have anti-aliasing filters to eliminate the effects of aliasing. The sampling rate of the chip is 26 kbs and can be lowered in software. The reference voltage is normally fixed at 2.42 V according to the reference circuit [1]. The chip has internal registers which can be accessed through SPI bus. There are

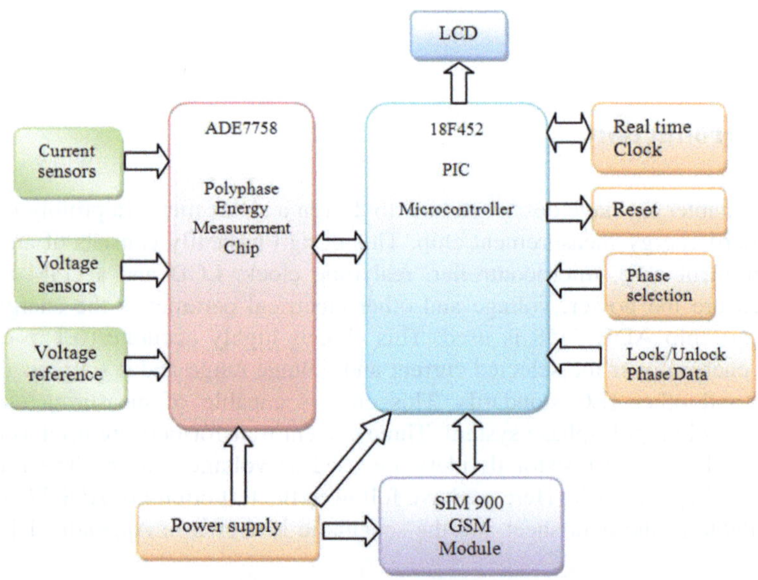

Fig. 4.1 Block diagram of the digital meter

three signed 24 bit voltage registers and current registers. The accuracy is 0.5 % within the defined range for each voltage and current input. The voltage channel and current channel should be calibrated according to the specified instructions provided in ADE7758 datasheet. There are three 16 bit signed watt-hour registers for energy information. They directly give the active energy accumulated in each phase over the given time period. Therefore it is much easier to calculate the active power and total energy using the PIC microcontroller. The VAR-hour registers provide the reactive energy accumulated within the specified time while the VA-hour registers give the apparent energy information. By reading the 10 bit FREQ register, we can calculate the frequency of each phase. When the chip is calibrated according to the given instructions, we can display all phase information including voltage, current, active power, reactive power, apparent power, power factor and frequency. Readings from the chip are processed and displayed as phase parameters on the LCD. The user can select the desired phase by pushing the push button in the meter.

4.3 Current Sensing Unit

ADE7758 has six analog inputs divided into two channels: current and voltage. The current channel consists of three pairs of fully differential voltage inputs. IAP and IAN, IBP and IBN, and ICP and ICN. These fully differential voltage input pairs have a maximum differential signal of ±0.5 V. In this design we have used 2000:1 ratio current transformer of accuracy of 0.1 %. The rated primary current is 20 A and maximum primary current it can handle is 60 A. It has working range of frequencies from 20 to 400 Hz. The rated burden resistor is 100 Ω. A simple resistor divider is used to match the input to the current channel. The CT shows linearity within the range of 50 mA to 25 A. The example design is shown in the Fig. 4.2.

Here we have shown an example calculation of matching a current sensor to the analog input of the chip. Figure 4.3 shows the current channel input specifications according to the Ref. [1].

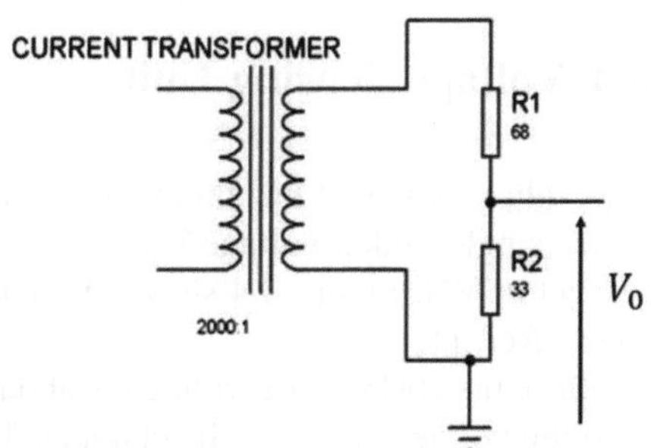

Fig. 4.2 Current transformer used for the current channel input

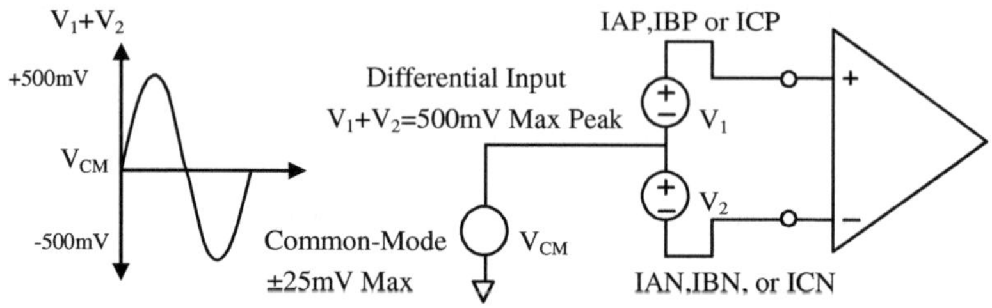

Fig. 4.3 Current channel input voltage range (Courtesy of Analog Devices, Inc.)

Consider the 2000:1 ratio current transformer used for the design. The rated primary current is 20 A and the burden resistor used at secondary is 100 Ω.

Output voltage at 100 Ω burden (at rated primary current) = $\frac{20}{2000} \times 100 = 1\,\text{V}$.

Maximum RMS voltage across channel=$0.5/\sqrt{2}$ =0.3535 V

$R1 + R2 = 100\,\Omega$ (since the burden is 100 Ω)

Therefore;

$$\frac{R2}{R1 + R2} \times 1 = 0.3535$$

$$\frac{R2}{100} \times 1 = 0.3535$$

$$R2 = 35.36\,\Omega$$

35.36 Ω is not available in the market. Available options are 33 and 68 Ω.

If we use above combination for $R1$ and $R2$, we can check for the maximum voltage across the channel at 20 A rated current.

Maximum voltage across R2 when 20 A primary current $= \frac{20}{2000} \times 33 \times \sqrt{2} = 0.46669\,\text{V}$

Maximum primary current for the saturation of the channel can be found vice versa.

Primary current for the saturation of the channel is 21.427 A. Therefore at 20 A rated primary current, the cannel is still below the saturation voltage.

4.4 Voltage Sensing Unit

The voltage channel has three single-ended voltage inputs: VAP, VBP, and VCP. These single-ended voltage inputs have a maximum input voltage of ±0.5 V with respect to VN. Figure 4.4 shows the voltage channel input specification according to the Ref. [1].

Since the RMS phase voltage is about 230 V, a simple resistor divider is used to feed the voltage signal to the channel. The values of the two resistors are 1 kΩ and

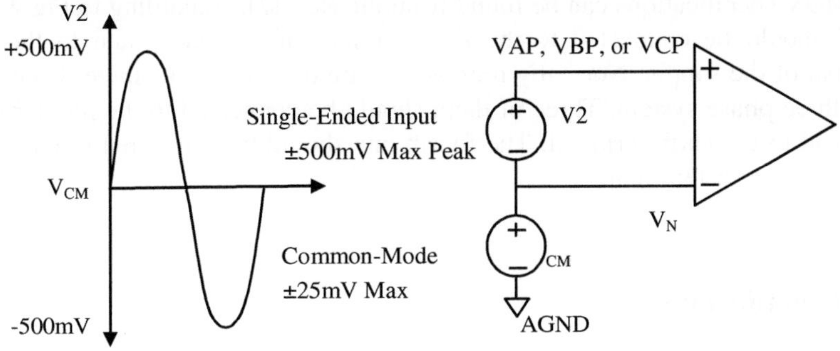

Fig. 4.4 Voltage channel input voltage range (Courtesy of Analog Devices Inc.)

1 MΩ. The power rating should be higher than 0.25 W. These values are chosen according to the reference design of ADE7758 [1]. The voltage sensing unit is shown in Fig. 4.5. V_{in} is the supply voltage (230 V). V_{out} is the output voltage which is fed to the voltage input channel of the energy chip.

Metal Oxide Varistor (MOV) is used for surge protection which is an essential component in all smart meter design. The MOV is a voltage-dependant resistor. Its resistance decreases as the voltage increases. The MOV is connected in parallel with the equipment which requires protection. In our case it is the voltage sensing unit. When there is a situation of an overvoltage, it forms a low resistance shunt and prevents any rise in voltage in the voltage sensing unit. For the design, MOV (V180MA3B) is selected. The peak surge current is 100 A (impulse of 8/20 μs). The maximum clamping voltage is 290 V. Peak energy is 700 mJ (10/1000 μs).

Fig. 4.5 Voltage sensing unit

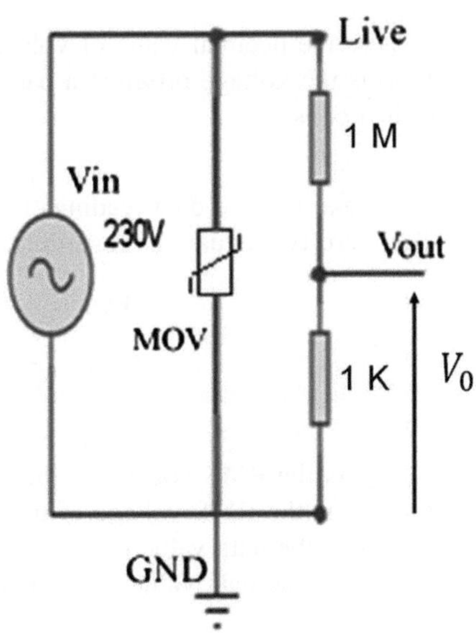

The MOV specifications can be found from the Ref. [2]. According to Fig 4.5, the MOV should be connected to the live terminal of the phase and to the earth terminal of the supply. Normally four MOVs are used when the meter is designed for a three phase system. Three of them should be connected to the live terminals each and to the earth terminal. The fourth one should be connected to the neutral and to the earth terminal.

4.5 Calculations

4.5.1 RMS Voltage and RMS Current Calculation

To calculate the RMS current and voltage, ADE 7758 chip has used methods explained in Sect. 3.4.1 (A) and (B) respectively. The RMS value of current and voltage signals can be obtained reading the IRMS register and the VRMS register on each phase. Voltage and current RMS values have an offset. This offset value can be calculated by following the instructions in calibration section in the data sheet [1]. When the offset values are known, the actual RMS value can be obtained. The following example shows how to clear the offset and get the actual RMS value of voltage and current using the data in RMS registers.

(A) Voltage offset calculation
The RMS voltage can be written as

$$V_{RMS} = k_v V_{DAT} + V_{OFS} \tag{4.1}$$

where;

V_{DAT} is the decimal value of voltage registry data
V_{OFS} is the voltage offset of a particular voltage channel
k_v is a constant.

Let's take voltage data readings from one channel at 25 and 230 V.
There are two equations regarding the RMS voltage

$$V_{RMS25} = k_v V_{DAT25} + V_{OFS} \tag{4.2}$$

$$V_{RMS230} = k_v V_{DAT230} + V_{OFS} \tag{4.3}$$

where;

V_{RMS25} is the RMS voltage of the reference meter at 25 V.
V_{RMS230} is the RMS voltage of the reference meter at 230 V.
V_{DAT25} is the data value of the registry at 25 V.
V_{DAT230} is the data value of the registry at 230 V.

By solving Eqs. (4.2) and (4.3) we can find V_{OFS} and k_v.

$$V_{OFS} = \frac{V_{DAT230}V_{RMS25} - V_{DAT25}V_{RMS230}}{V_{DAT230} - V_{DAT25}} \quad (4.4)$$

$$k_v = \frac{V_{RMS230} - V_{RMS25}}{V_{DAT230} - V_{DAT25}} \quad (4.5)$$

Therefore it is needed to get two data readings at 25 and 230 V. By using Eqs. (4.4) and (4.5) one can calculate the V_{OFS} and k_v for each phase. This should be done by connecting the design with a reference voltage meter. However the calibration process has several additional methods. In the data sheet it has been recommended that the voltage readings should be taken synchronous to zero crossings of each voltage channel. It can be done by setting the interrupts of the ADE chip to zero crossing by selecting a particular phase. Further in the data sheet it has been mentioned that there should be 20 readings to get an average value. This method will lead to more accurate voltage measurement rather than getting two data sets from the chip. More details can be found from calibration sector of the chip [1].

(B) Current offset calculation

The current channel also has an offset which should be found to get the actual RMS current. The RMS current variation can be written with respect to the offset by (4.6).

$$I_{RMS}^2 = k_i I_{DAT}^2 + I_{OFS} \quad (4.6)$$

where;

I_{DAT} is the decimal value of current registry data
I_{OFS} is the current offset of a particular current channel
k_i is a constant.

Let's take the current data readings from one channel at 40 mA and 10 A. 40 mA is 1/500 of the full scale current in our design.

The RMS current can be written as:

$$I_{RMSm}^2 = k_i I_{DATm}^2 + I_{OFS} \quad (4.7)$$

$$I_{RMSt}^2 = k_i I_{DATt}^2 + I_{OFS} \quad (4.8)$$

where;

$I_{RMSm}$is the reference meter current at 40 mA.
$I_{RMSt}$is the reference meter current at 10 A.
I_{DATm} is the data value of registry at 40 mA.
$I_{DATt}$is the data value of registry at 10 A.

By solving Eqs. (4.7) and (4.8) one gets:

$$I_{OFS} = \frac{I^2_{RMSm} I^2_{DATt} - I^2_{RMSt} I^2_{DATm}}{I^2_{DATt} - I^2_{DATm}} \tag{4.9}$$

$$k_i = \frac{I^2_{RMSt} - I^2_{RMSm}}{I^2_{DATt} - I^2_{DATm}} \tag{4.10}$$

To find I_{OFS} and k_i values, 40 mA and 10 A currents should be fed to the system. Here we have shown only the equations that can be used to remove the offset of current channel. These equations are very much helpful when deriving the RMS current by external MCU. However to get more accurate RMS reading, the data values of the registers should be taken synchronously with the zero crossings of the current signal. More details about calibration can be found from the calibration sector of the ADE7758 data sheet [1].

(C) RMS Voltage Measurement

RMS voltage can be obtained by reading the RMS registers in the chip. There are three RMS voltage registers AVRMS, BVRMS and CVRMS for each phase. RMS voltage shows linearity over full scale voltage to 1/20 of the full voltage. Rated voltage of the design is around 230 V. Therefore a good accuracy is obtained between 23 V and 230 V. The measurement error is 0.5% within this limit. When the voltage is lower than 10 V, there is some noise which cannot be avoided.

Figure 4.6 shows the RMS registry value variation in phase A against the reference meter RMS voltage. The data are obtained from the AVRMS register. This is a 24 bit signed register. The binary value is converted to decimal in the MCU. The RMS voltage is measured using a reference meter (of accuracy 0.1 %) and the voltage is fed with an adjustable voltage source.

Figure 4.7 shows the RMS voltage with respect to the reference meter after calibration.

The results obtained for other phases have the same pattern including linearity and accuracy of RMS voltage.

(D) RMS Current Measurement

There are three registers for RMS current measurement in the energy chip. They are AIRMS, BIRMS, CIRMS for phase A, phase B and phase C respectively. They are 24 bit readable signed registers. The values shown in these registers are linear over full scale to 1/500 full-scale value [1]. In this study, we have chosen 21.427 A as full scale current. Therefore we obtain a linearity for RMS current from 21.427 A to 40 mA. The accuracy is 0.5 % within this range.

Figure 4.8 shows the output from AIRMS register for different RMS currents before calibration. The binary value obtained from the register is converted to decimal before the graph is plotted. The current is measured using a reference meter of accuracy 0.1 %.

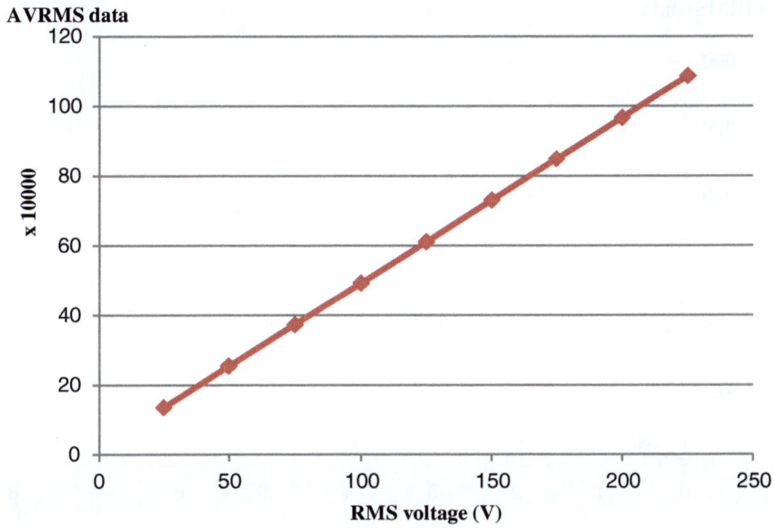

Fig. 4.6 AVRMS data variation against the reference meter voltage

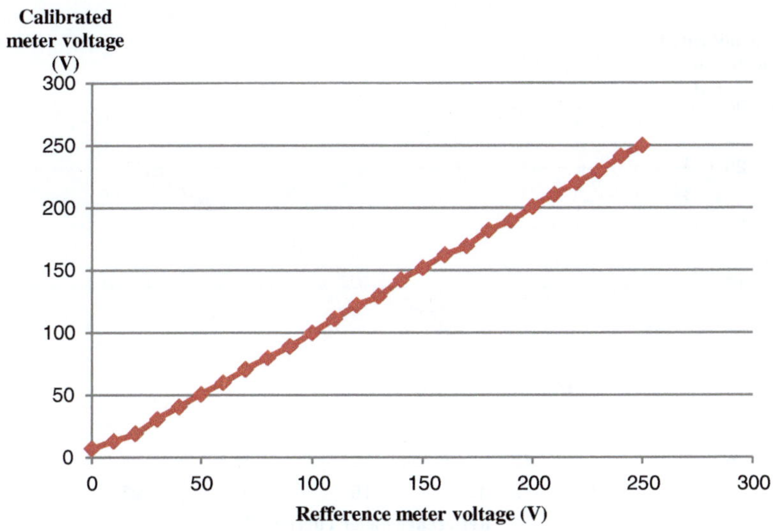

Fig. 4.7 RMS voltage variation with respect to the reference meter

The current cannel has an offset which should be corrected by proper calibration. We could obtain the RMS current variation with the desired accuracy of 0.5 % after the calibration.

Figure 4.9 shows the RMS current variation against the reference meter after the calibration.

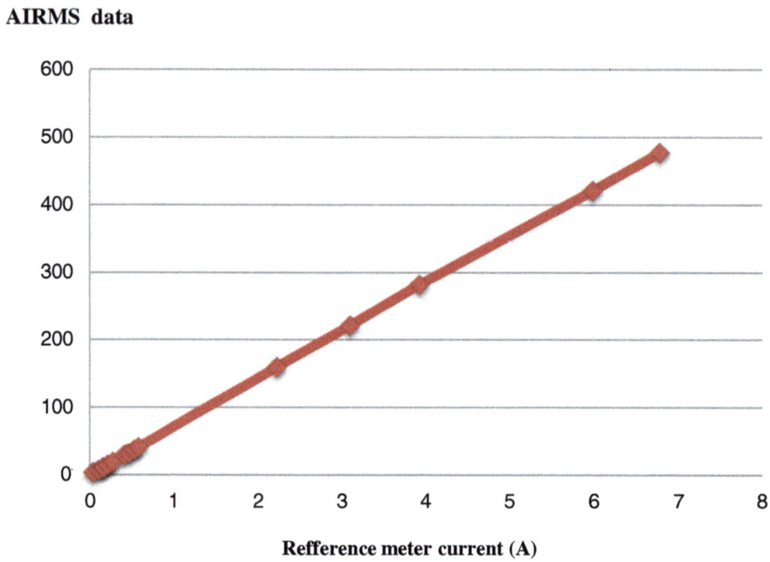

Fig. 4.8 AIRMS data variation against the reference meter current

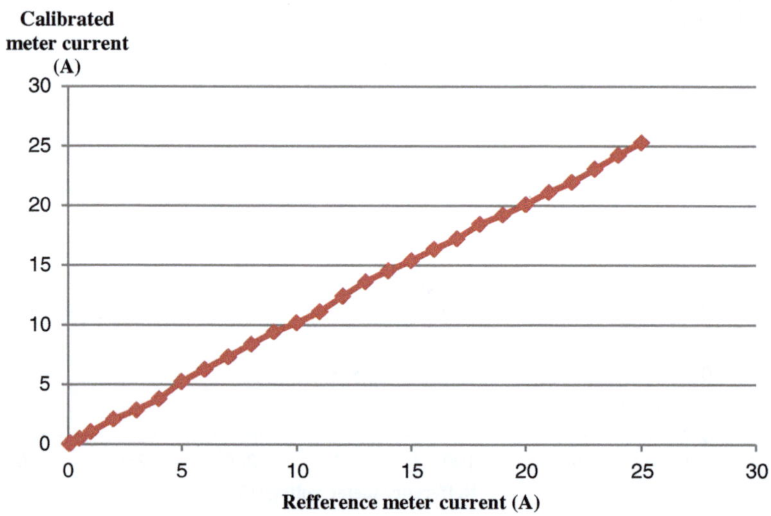

Fig. 4.9 RMS current variation against the reference meter

The results obtained for other phases have the same pattern including linearity and accuracy for RMS current.

4.5.2 Active Power and Energy Calculation

Active power calculation is done according to the method described in the Sect. 3.4.2 (A) of the ADE chips. The active power of each phase accumulates in the corresponding 16-bit watt-hour register (AWATTHR, BWATTHR, or CWATTHR). Those registers can overflow if the reading time is large. Therefore it is necessary to study the operation of these registers by referring to the data sheet. If we consider the register access time as t_0

$$T \leq t_0 \leq T_r \tag{4.11}$$

where;

T is the line cycle time
(for 50 Hz system T = 20 ms and for 60 Hz system T= 16.668 ms)
T_r is the registry over flow time

The registers can be read at t_0 by setting the timer interrupts in the MCU. If the registry access time (t_0) and energy accumulated in those registers are known, one can easily calculate the average power within the period. Even though setting a timer interrupt makes known access time, it has a disadvantage when calculating the total apparent power which will be discussed later. The other recommended method is the reading of these registers synchronously to zero crossing of each phase. This can be done by setting the interrupt registers in the ADE 7758 chip to generate interrupts at zero crossings. The interrupts can be caught by connecting the IRQ pin in ADE7758 to one of the interrupt pins in the MCU. In this method the line cycle time period should be known to calculate the power in each register. The line cycle time period can be found by reading the FREQ register in ADE 7758.

Active power also contains an offset which should be eliminated from the calibration process. The ADE7758 incorporates a watt offset register on each phase (AWATTOS, BWATTOS, and CWATTOS). These are signed two's complement, 12-bit registers that are used to remove offsets in the active power calculations.

The chip continuously accumulates the active power signal in the internal 41-bit energy registers. The user can read the watt-hour registers (AWATTHR, BWATTHR, and CWATTHR) which represent the upper 16 bits of these internal registers. WATTHR registers give the energy accumulated over a specified time. When the register is read, the data in the register are cleared. Therefore the active power and the active energy are calculated by reading the data from register. When the data are read, (4.12) is used for active power calculation in external MCU.

$$\text{Active power} = \mathrm{k_a}\frac{\text{registry value}}{t_0} + \mathrm{C_a} \tag{4.12}$$

where;

k_a is the constant for active power and C_a is the offset.

These values can be found in the calibration process [1]. Active energy calculation is done in the MCU by following the (4.13).

$$\text{Active energy} = \sum \text{active power} \ \times t_0 \tag{4.13}$$

This energy can be converted to watt-hour by using a software.

Figure 4.10 shows the AWATTHR data variation with the reference meter active power. The values are obtained from registers synchronized to interrupts generated by the MCU at every $t_0 = 0.52$ s intervals.

Active power and active energy for other phases can be found using the same procedure discussed above. They have the same linearly pattern as in the Fig. 4.10.

Finally the total active power can be found using following equation.

$$Total\,active\,power = P_a + P_b + P_c \tag{4.14}$$

where;

P_a, P_b and P_c are active power of each phase a, b, and c respectively.

And the total active energy is given by

$$Total\ active\ energy = E_{pa} + E_{pb} + E_{pc} \tag{4.15}$$

where;

E_{pa}, E_{pb} and E_{pc} are active energy of each phase a, b and c respectively.

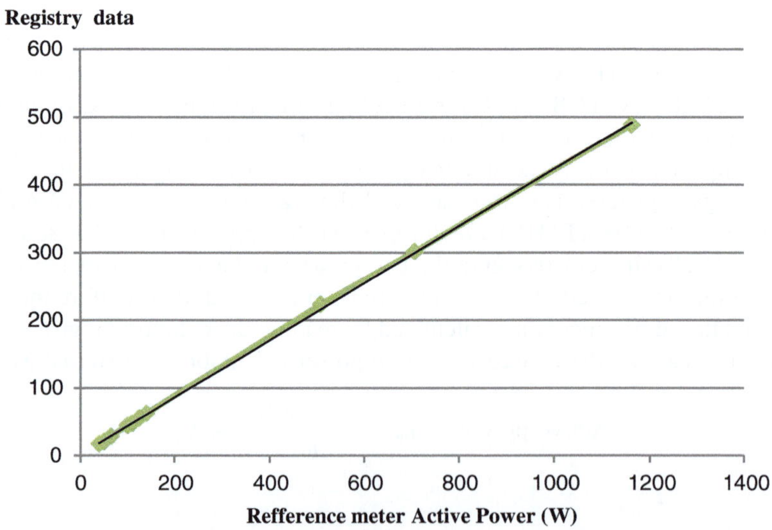

Fig. 4.10 AWATTHR data variation with the reference meter active power

4.5.3 Reactive Power and Energy Calculation

Reactive energy calculation is done as same as the method used in ADE chips which has been described in Sect. 3.4.3. Even the reactive power calculation can contain an offset which can be removed by writing the values to AVAROS, BVAROS, and CVAROS 12 bit registers by proper calibration.

To calculate the reactive power, the following equation can be used in the MCU.

$$\text{Reactive power } = \mathrm{k_r}\frac{\text{registry value}}{t_0} + \mathrm{C_r} \tag{4.16}$$

where;

$\mathrm{k_r}$ is the constant for reactive power and $\mathrm{C_r}$ is the offset

These values can be found in the calibration process [1]. Reactive energy calculation is done in the MCU by following (4.17).

$$\text{Reactive energy} = \sum \text{Reactive power } \times t_0 \tag{4.17}$$

For a three phase system the total reactive power can be written as

$$\text{Total reactive power} = Q_a + Q_b + Q_c \tag{4.18}$$

where;

Q_a, Q_b and Q_c are reactive power of each phase a, b, and c respectively.

Reactive energy is defined as the integral of reactive power. Reactive energy of each phase is accumulated in the corresponding 16-bit VAR-hour register (AVARHR, BVARHR, or CVARHR). With the MCU we can directly read the reactive energy accumulated within the desired time period. The registers should be read and cleared before they overflow. This can be adjusted in the software. Reactive energy can be accumulated at the MCU and converted to kvarh.

The total reactive energy is given by:

$$\text{Total reactive energy} = E_{qa} + E_{qb} + E_{qc} \tag{4.19}$$

where;

E_{qa}, E_{qb} and E_{qc} are reactive energy of each phase a, b and c respectively.

4.5.4 Apparent Power Calculation

$$\text{Apparent power } = V \times I \tag{4.20}$$

Apparent power is obtained by multiplying the RMS value of the current and the voltage in each phase. The output from the multiplier is then low-pass filtered to obtain the average apparent power. Each RMS measurement includes an offset compensation register to calibrate and eliminate the dc component in the RMS value. The offset compensation of the apparent power measurement in each phase should be done by calibrating each individual RMS measurement. There are three voltage RMS registers and three current RMS registers. The offset value should be found from the calibration process and written to the RMS offset registers. Then the values from the RMS registers can be used to calculate and display the apparent power.

However for a three phase sinusoidal unbalanced system one cannot add apparent power in each phase unless they have the same power factor. To get the vector apparent power of unbalanced systems, the vector sum of the active and reactive powers should be found. This can be illustrated using a power triangle as shown in Fig. 4.11.

To get the vector apparent power following equation can be used [3]

$$S_V = \sqrt{P_T^2 + Q_T^2} \tag{4.21}$$

where;

S_V is the vector apparent power
P_T is the total active power in one line cycle
Q_T is the total reactive power in one line cycle
θ is the equivalent phase angle for an unbalanced system

However P_T and Q_T values should be found synchronized to zero crossings of each phase. For an example if the ADE chip is programmed to generate interrupts at every zero crossing of each phase, one can easily calculate the P_T and Q_T .

Considering one cycle in each phase:

$$P_T = P_{aT} + P_{bT} + P_{cT} \tag{4.22}$$

$$Q_T = Q_{aT} + Q_{bT} + Q_{cT} \tag{4.23}$$

where;

P_{aT}, P_{bT}, P_{cT} are active power in each phase a, b and c respectively for one line cycle

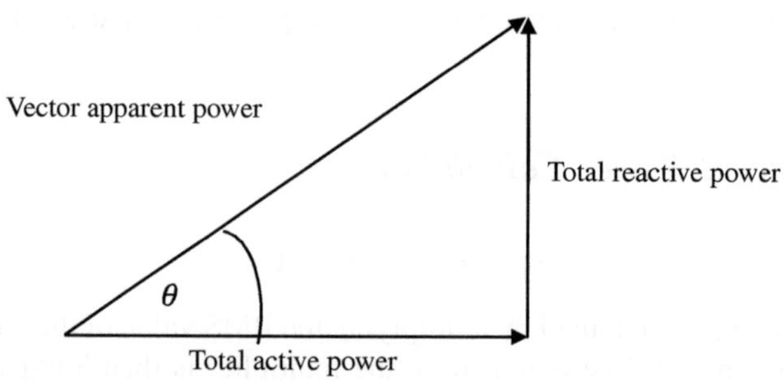

Fig. 4.11 The power triangle

Q_{aT}, Q_{bT}, Q_{cT} are reactive power in each phase a, b and c respectively for one line cycle

Apparent energy can also be found in the MCU using equation (4.24)

$$\text{Total apparent energy} = \sum S_V.T \tag{4.24}$$

where;

T is the line cycle period.

4.5.5 Power Factor Calculation

The power factor calculation is done inside the MCU. Power factor is defined as the ratio between the active power and the apparent power. It can be calculated for one particular phase in a three phase system. When the load is balanced, the power factor in one phase is equal to the power factors in other two phases. However when there is an unbalanced load, the power factor in each phase is different. Since we are reading the active power and reactive power in each phase synchronously to zero crossings, we can calculate the equivalent power factor for the three phase system whether it is balanced or unbalanced according to the Eq. (4.25).

$$P.F = \frac{pa + pb + pc}{\sqrt{(pa + pb + pc)^2 + (qa + qb + qc)^2}} \tag{4.25}$$

where;

pa, pb, pc are the active powers and *qa, qb, qc* are the reactive powers in each phase a, b and c respectively

$P.F$ is referred to equivalent power factor for a three phase system [4].

4.5.6 Frequency Calculation

The ADE7758 contains three different circuits for zero crossing detection in each voltage channel (VAN, VBN, and VCN). The interrupt signal is fed from the energy chip to the MCU in order to read the frequency or the line period. This information is updated every four periods of the selected phase in the 12-bit FREQ register. The user can select one of the voltage channels for frequency measurement by writing to the MMODE register. Bit 7 of the LCYCMODE selects whether the period register provides the frequency or the period. Setting this bit causes the register to display the period. The default setting is logic low, which causes the register to display the frequency.

4.5.7 Power Quality Measurements

Line sag detection, peak current and peak voltage detection in each phase are recorded within 15 min. When the SAGLVL register is set to user specified value (162 V) and SAGCYC register is set to 6 half cycles, the SAG detection is active. If the SAG flag of corresponding phase in the interrupt status register is set, SAG event is recorded at the end of the sixth half cycle. This event can be caught from the interrupt registers within the chip [1]. After setting the VPINTLVL register to the full scale value and the required phase in the LCYCMODE register, interrupts are generated when the voltage exceeds the defined value. Therefore, choosing phase A as the voltage channel and the full scale voltage as 290 V, we can activate the peak voltage detection in phase A. Similarly, for the peak current detection, the IPINTLVL register can be set to a required value, 21.427 A in this case.

4.6 Energy Chip and MCU Communication

The energy chip has a built-in SPI interface [1]. Therefore in this design, the serial interface is used with the MCU for communication. The pin diagram and other specifications on PIC 18F452 microcontroller can be found in [5]. The functionality of ADE7758 is accessible via several on-chip registers. The contents of these registers can be updated or read using the on-chip serial interface. Figure 4.12 shows the pin configuration of the energy chip and the MCU. A common ground should be provided for the energy chip and the MCU. A pull up resister should be connected to the *IRQ* pin. A digital port can be assigned for SPI communication in the PIC18F452. The PORTB is used for SPI. The complete schematic is shown in Appendix 2.

4.7 Power Supply Unit

A continuous power supply is essential for the operation of the digital meter. A 230 V/12 V step-down transformer with maximum 1 A secondary current is used to provide the required power. Three DC voltage levels are used (5 V, 3.8 V, and 3.3 V). This is because the PIC microcontroller operates at 5 V, the GSM module operates at 3.8 V, and the RTC operates at 3.3 V. The LM7805 regulator IC is used for the 5 V supply and LD33 V is used for the 3.3 V supply. The GSM module operates between 3.2 to 4.8 V. Therefore a single diode is connected to the output of the LM7805 to provide a voltage of about 4.3 V. The schematic of the power supply is shown in Appendix 4. A battery backup is used in addition to the step–down transformer power unit. This circuit charges a 12 V rechargeable battery when the power is present. When the power is off this battery can handle all

Fig. 4.12 Pin configuration (ADE7758 and 18F452)

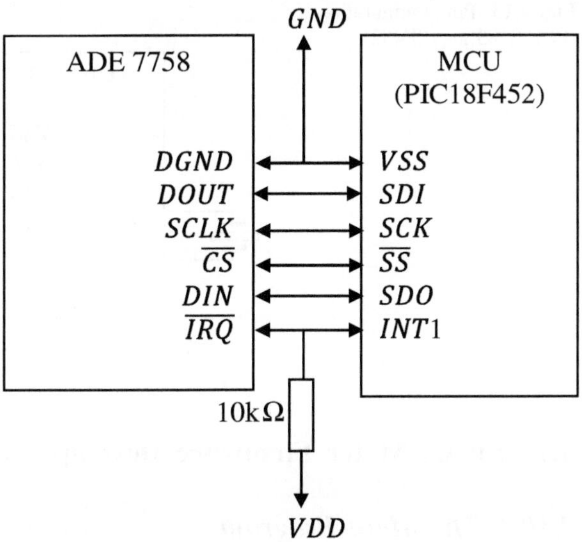

the components in the smart meter for at least 24 h. Thus the PIC microcontroller goes into the sleep mode when the power is off in the main supply. The PIC microcontroller writes the last data obtained from the energy chip to its EEPROM before it goes into the sleep mode.

4.8 Data Backup

The EEPROM available in the PIC microcontroller is used to store the active energy, reactive energy and maximum demand. Active energy consumed and active energy exported are separately stored in the EEPROM whenever they are incremented by 0.1 kWh. In the event of a power failure or, a watch dog reset and subsequent recovery, the previously accumulated energy is retrieved and accumulation resumes from the last value.

4.9 Real Time Clock

The RTC is driven by a LD33 V regulator IC with a 3.3 V output. The PCF 8583 clock calendar chip is interfaced to the PIC microcontroller using two wire serial interfaces [6]. Here the I2C is used as the communication protocol. The *OSCI* and *OSCO* pins should be connected to a 32.768 oscillator. The *SCL* and *SDA* pins should be pulled-up with resistors and a common ground should be provided. Figure 4.13 shows the pin configuration of the RTC and the MCU. The complete schematic can be viewed in Appendix 2.

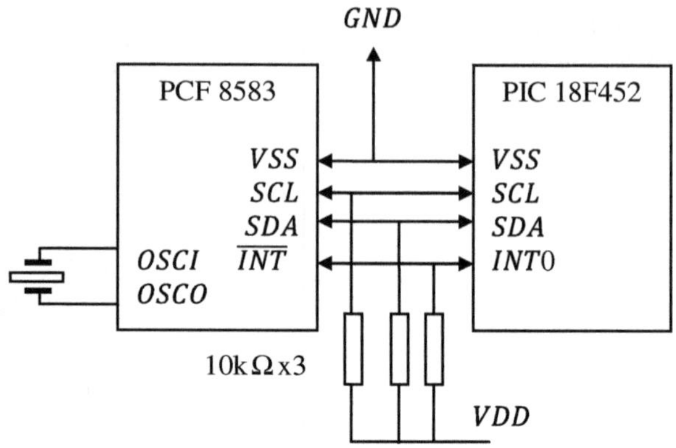

Fig. 4.13 Pin configuration (PCF 8583 an 18F452)

4.10 Smart Meter Firmware Development

4.10.1 The Main Program

The PIC microcontroller is initiated in the main program. This includes the settings of the LCD, the RTC, the watch dog timer, the interrupts and the serial port. The LCD display is connected to PORT B in the microcontroller and initiated accordingly. The RTC is connected to PORT C and the software I2C communication is enabled. The interrupts are enabled as timers and external interrupts. The timer interrupt has been selected to generate interrupts every 0.52 s intervals. Therefore t_0 becomes 0.52 s. Here we haven't enabled the interrupt in ADE7758 because of the fixed access time t_0 .When the microcontroller generates the Timer0 interrupt, the readings are taken from the energy chip. The external interrupt is enabled to catch the falling edges where the signal comes to the INT0 interrupt pin in the PORT B. These interrupts are generated by the RTC as alarm signals. Interrupt priority is enabled by setting the IPEN bit in the RCON register. The serial port is connected to the GSM module which gets signal from the microcontroller as AT commands. Hence the baud rates must be set to match. The energy chip communicates with the microcontroller via SPI bus.

Initially the program does not read any information from the EEPROM. However in the long run it stores the active energy, reactive energy and maximum demand in the EEPROM. When the program gets stuck, the watchdog time out occurs and the program will restart. In such events it reads the previously stored data from the EEPROM and takes those readings for the calculations.

In the next step it initiates the energy chip by writing specific values to the internal registers. This includes the enablement of the active and reactive power pulse out puts. The calibrated data are also written to the gain and offset registers in the chip.

The real time clock initialization is done next. This includes setting the date, time and the timer alarm as 1 min. This process is done only for one time and it will never run again during a reset. Once it is initiated, the RTC runs with the continuous power from the Lithium–ion 3 V battery.

After the RTC is initialized, it goes to the infinite loop, where it runs inside the while loop which is true always, and continuously checks for the active and reactive energy. When the accumulated active or reactive energy equals or exceeds 0.1 kWh or 0.1 kvar, it stores the total active and reactive energy information in the EEPROM at regular intervals. When this is done it looks for any available data packets generated in the interrupt section. Data packets are generated every 15 min. If any data packet is available it sends the data via SMS by writing the AT commands to the GSM module. Otherwise it goes back to the start of the while loop. The algorithm of the main program is shown in Fig. 4.14. The software can be prepared with Ref. [5] and using a language that supports PIC programming.

4.10.2 TIMER0 Interrupt Routine

The timer interrupt is selected as TIMER0 interrupt within the PIC microcontroller. It is configured to generate interrupts in every 0.52 s. When an interrupt event occurs, the microcontroller gives priority to the instructions in this interrupt section. First step is to read the data from the energy chip. This includes reading voltage and current RMS values from RMS registers, reading active and reactive energy accumulated in the energy registers, frequency from the frequency register and interrupt status register.

All the data come from the registers are in binary and should be converted to decimal for calculations. In the calculation process it clears the offsets and sets the gain for voltage and current RMS for each phase. The active and reactive energy accumulated within the interrupt time period is proportional to the active and reactive power. Therefore the energy values are multiplied from the gain values to obtain the active and reactive power.

Then per phase and three phase parameters are calculated according to the calculations done in the calibration section. Per phase information includes RMS voltage, RMS current, apparent power, active power, reactive power, power factor, frequency, active energy inflow and active energy outflow for phase A, phase B and phase C. Three phase information includes line to line voltages, total active power, total reactive power, total apparent power, equivalent power factor, active energy inflow, active energy outflow, total reactive energy and average demand.

The information to be shown on the LCD can be selected using the push button. The default position shows the phase A information and by pushing the button the user can view the phase B, phase C, three phase, time and date information. When it goes to a particular phase, it automatically rotates the phase information such as voltage, current, and active power. To stop the run mode and select the required

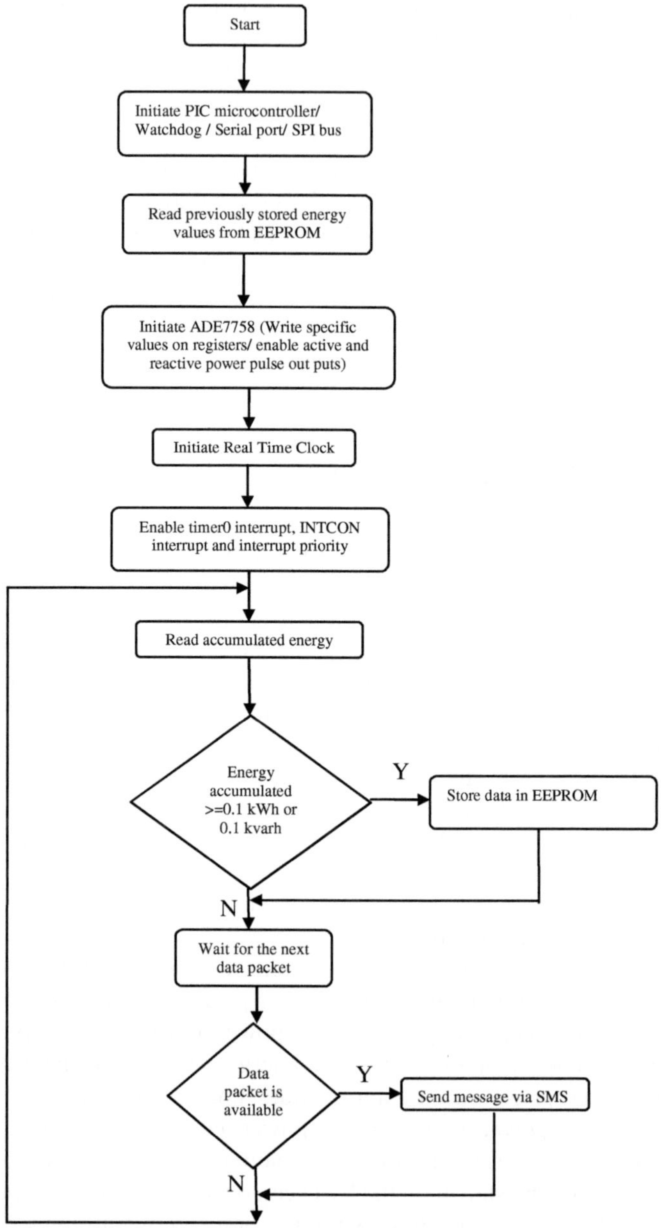

Fig. 4.14 Smart meter firmware (main program loop)

phase information, the user can adjust the run/stop button. The TIMER0 interrupt looks for the user inputs from those two buttons and selects the required phase and required phase quantity to be shown on the LCD.

Then it checks for the accumulated energy values. If the total active or reactive energy consumption exceeds 0.1 kWh or 0.1 kvar, it updates a variable in the main program where the data are stored in EEPROM. Finally it returns from the interrupt routine by clearing the timer0 flag. Figure 4.15 shows the energy measurement TIMER0 interrupt routine.

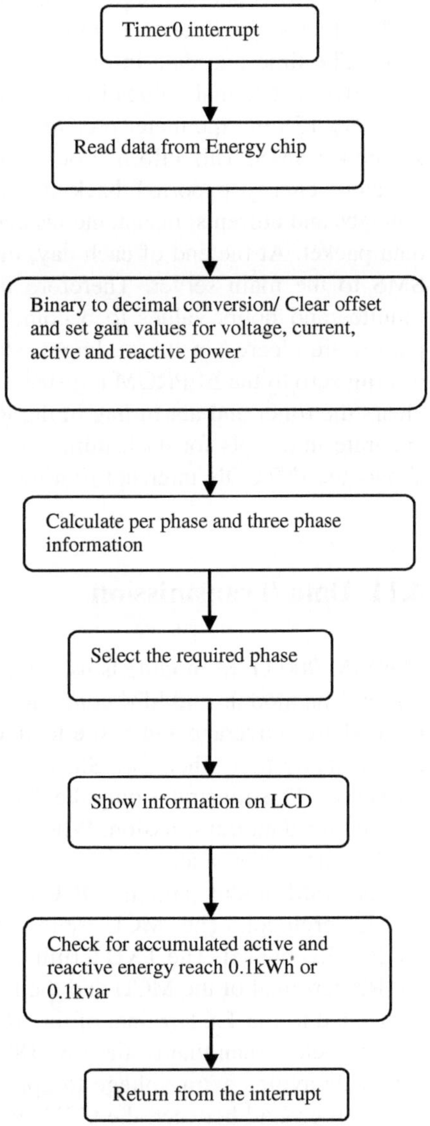

Fig. 4.15 Smart meter firmware (TIMER0 interrupt routine)

4.10.3 INTCON Interrupts Routine

The INTCON interrupt is enabled to handle the instructions due to external interrupts. The RB0/INT0 pin in the PIC microcontroller is connected to the open drain interrupt output in the RTC. When an interrupt is generated by the RTC due to a timer alarm, it will be detected by the PIC microcontroller. The RTC is configured to generate interrupts in every minute, in real time.

The first step in the INTCON interrupt is to read the time and date from the RTC. The time and date information are binary values. These binary values are converted to decimal values in the next step.

Every 15 min, the meter prepares per phase information to be sent to the main server via SMS. This information includes total active and reactive energy accumulated, energy exported back to the system, frequency, instantaneous RMS voltages and currents, instantaneous active and reactive power, and the time of the data packet. At the end of each day, the power quality measurements are sent via SMS to the main server. Therefore in the interrupt routine, it checks for both minutes and hours values to become zero. The active, reactive energy in each phases are cleared in the midnight on 26th day of each month. This is done by writing zero to the EEPROM registers. Before it exists from the interrupt routine, it clears the timer and alarm flag in the RTC. This process must be done in order to generate interrupts for each minute synchronized with the real time. Figure 4.16 shows the INTCON interrupt routine.

4.11 Data Transmission

The SIM 900 GSM module is used for communication between the meter and the server. The module and SIM configurations are done according to the Ref. [7]. Due to good area coverage and cost effectiveness, sending data via SMS turns out to be a very handy tool. Therefore SMS is preferred to transmit data from the meter to the server. This illustrates only the data formats and configurations. GPRS can also be used for data transmission. When GPRS is selected, a socket listener should be used at the server side.

The GSM module and the MCU communicate with each other according to the RS232 protocol. The MCU sends AT commands to the GSM module via the RS232 channel. The TXD terminal of the GSM module should be connected to the RX terminal of the MCU and the RXD terminal of the GSM module should be connected to the TX terminal of the MCU. Normally, the GSM module operates a voltage below than that of the PIC 18F452. Adding a 1 kΩ resister between these terminals avoids extra voltage to appear at the GSM module. There should be a common ground between the GSM module and the MCU. Figure 4.17 shows the pin configuration of the MCU and the GSM module. The complete schematic can be found in Appendix 3.

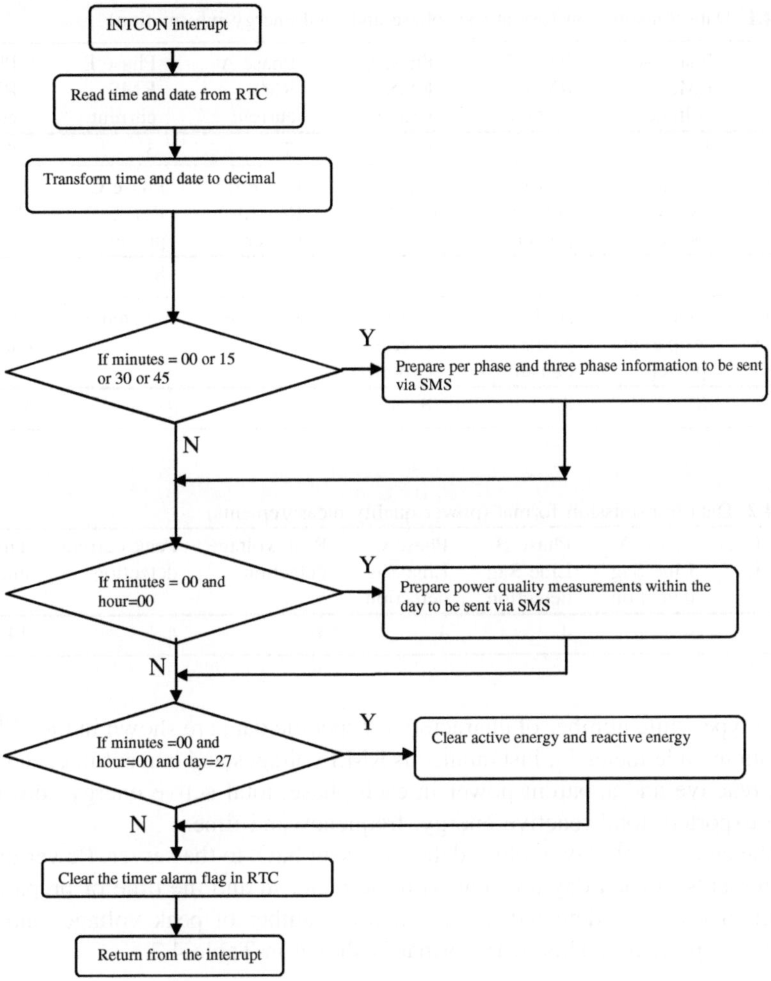

Fig. 4.16 Smart meter firmware (INTCON interrupt routine)

Fig. 4.17 Pin configuration (GSM module and MCU)

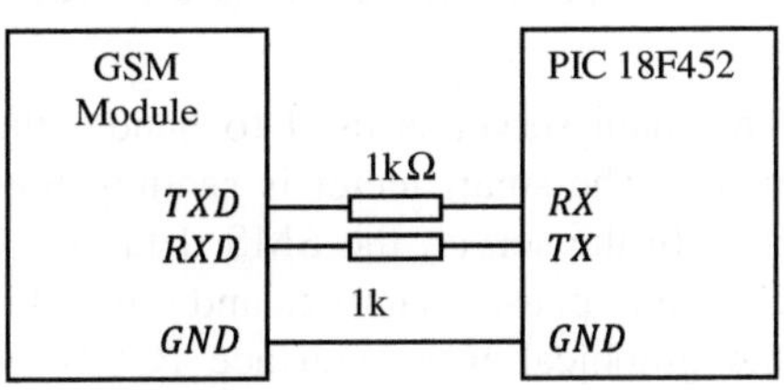

As described in the interrupt section we are using three SMS formats to transmit the data. In every 15 min the meter prepares per phase information to be sent to the main server. The first data format is shown in the Table 4.1. The first row shows

Table 4.1 Data transmission format (per phase and total energy information)

Meter Id	Phase A RMS voltage	Phase B RMS voltage	Phase C RMS voltage	Phase A RMS current	Phase B RMS current	Phase C RMS current
4	5	5	5	5	5	5
Phase A Active power	Phase B Active power	Phase C Active power	Phase A Reactive power	Phase B Reactive power	Phase C Reactive power	Phase A Apparent power
8	8	8	8	8	8	8
Phase B Apparent power	Phase C Apparent power	Active energy consumed	Active energy exported	Reactive energy	Frequency	Time and date
8	8	8	8	8	4	12

Table 4.2 Data transmission format (power quality measurements)

Meter Id	Data Id	Phase A Line Sag detection	Phase B Line Sag detection	Phase C Line Sag detection	Peak voltage detection	Peak current detection	Time and date
4	2	4	4	4	4	4	14

the data type while number of characters for specific data are shown in the 2nd row. This data include meter Id, instantaneous RMS voltages and current in each phase, active, reactive and apparent power in each phase, total active energy consumed, energy exported, total reactive energy, frequency and time.

At the end of each day another data set is sent back to the server. Power quality measurements in each day are sent with the meter id and the time of preparation. Number of line sags detected in each phase, number of peak voltage, and peak currents are prepared. This SMS format is shown in Table 4.2.

4.12 The Serial Port Listener

The main server is used to handle the SMS which are coming from the smart meter. The smart meter is sending SMS in every 15 min and at the end of each day. In the server, the SMS data are separated and used for further calculations. Basically phase quantities and three phase quantities are extracted and displayed in the graphical user interface (GUI). Active energy information and maximum demand are used for energy cost calculations. The user can view his consumption details by sending a SMS to the server requesting the data.

A GSM module is connected to the computer via the serial port to collect the SMS coming from the meter side. The serial port is listened by a java program.

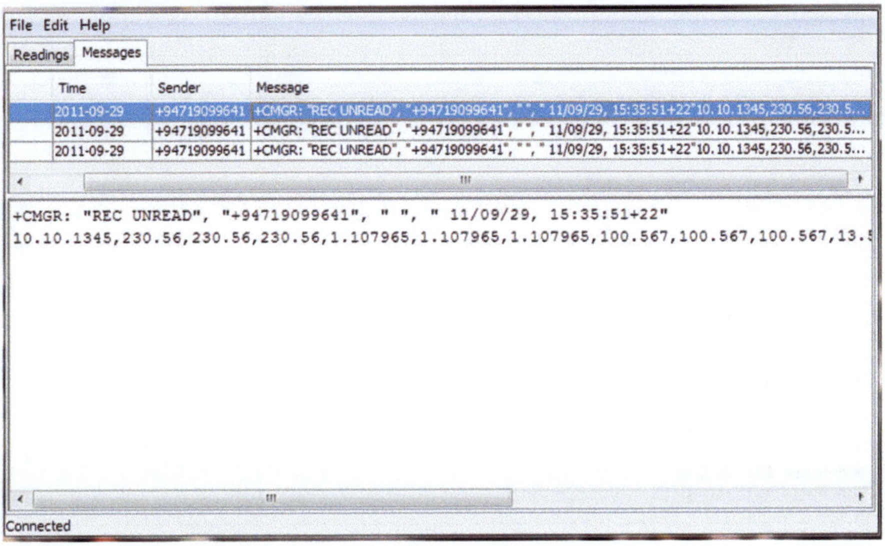

Fig. 4.18 Serial port listener

Whenever a new message comes to the GSM module, the program reads it by sending AT commands. All values are then separated by "," character. Figure 4.18 shows how the raw data are displayed using the port listener.

4.13 The Graphical User Interface

At the server side, a GUI is used to monitor the data coming from the smart meter. It displays the instantaneous RMS voltages, RMS current, apparent power and active power. The total energy consumed and its cost, energy exported and the credits earned are also calculated and displayed. The operator can view the historical data as well. Figure 4.19 shows a screen shot of the GUI.

4.14 Bill Generation, Cost Prediction and Customer Update

Electricity bill is generated considering the time of use tariff. Active energy usage, fixed charge, and maximum demand charge are taken while preparing the bill.

The 24 h have been divided according to the time of use and shown in Table 4.3.

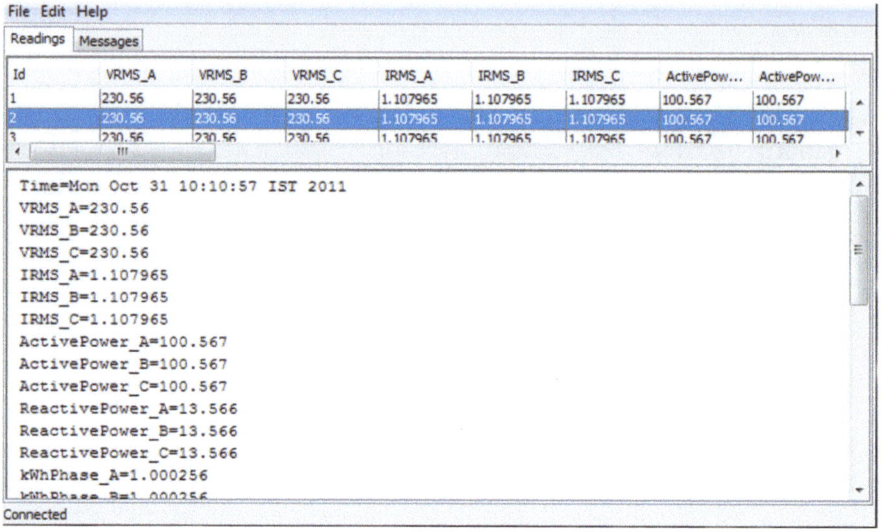

Fig. 4.19 The GUI at the server side

Table 4.3 Time of use

Session	Time (hrs)
Day	05:30–18:30
Peak	18:30–22:30
Off-peak	22:30–05:30

With the information received by the server, the current cost of energy consumption, estimated bill for the month and average daily energy consumption are calculated. When a particular consumer exports the energy back to the system, the credits available for him are also calculated.

The total cost (T) is given by

$$T = x_1 \sum E_1 + x_2 \sum E_2 + x_3 \sum E_3 + k_1(MD) + k_2 \tag{4.26}$$

where

k_1 is the maximum demand charge per unit (LKR/kVA).
k_2 is the fixed charge (LKR/month).
x_1 is the energy charge at day use (LKR/kWh).
x_2 is the energy charge at peak use (LKR/kWh).
x_3 is the energy charge at off-peak use (*LKR*/kWh).
E_1 is the Active energy usage during the day period (kWh).
E_2 is the Active energy usage during the peak period (kWh).
E_1 is the Active energy usage during the off-peak period (kWh).
MD is the maximum demand in the month (kVA).

The average daily energy consumption (E_{avg}) is given by

$$E_{avg} = \frac{\sum E_1 + \sum E_2 + \sum E_3}{t} \quad (4.27)$$

where t is the number of days the energy is consumed.

The average daily energy cost (C) is given by

$$C = \frac{x_1 \sum E_1 + x_2 \sum E_2 + x_3 \sum E_3}{t} \quad (4.28)$$

The cost prediction (P) for the month is given by

$$P = k_1(MD) + k_2 + nC \quad (4.29)$$

where;

n is the number of days for a particular month.

The consumer can check the number of days remaining by providing the amount that he or she would like to pay. The number of remaining days (N) is given by

$$N = \frac{A - k_1(MD_m) - k_2}{C} - t \quad (4.30)$$

where;

A is the user desired amount to pay for the month

$MD_m$is maximum demand for the day of consideration in the month.

4.15 Electricity Consumption Details

The meter was installed in a domestic premise and allowed to run continuously for 10 days. The active energy information received by the server was stored in a data base. Table 4.4 shows the data base value obtained.

$$\sum E_1 = 10.635\,\text{kWh}$$
$$\sum E_2 = 14.173\ \text{kWh}$$
$$\sum E_3 = 6.128\ \text{kWh}$$

By taking

k_1 = 0 LKR (for domestic consumers)

k_2 = 150 LKR

x_1 = 8.99 (LKR/kWh)

x_2 = 11.11(LKR/kWh)

x_3 = 8.06 (LKR/kWh)

Table 4.4 Energy consumption details under time of use

	Day usage (kWh)	Peak usage (kWh)	Off-Peak usage (kWh)
1	1.021	1.441	0.613
2	0.982	1.375	0.622
3	1.019	1.325	0.610
4	1.052	1.322	0.623
5	1.104	1.401	0.612
6	0.991	1.428	0.624
7	1.173	1.503	0.598
8	1.019	1.501	0.605
9	1.204	1.437	0.610
10	1.073	1.440	0.611

Average day session energy usage

$$= \frac{\sum E_1}{t} = \frac{10.635}{10} = 1.0635 \text{ kWh}/day$$

Average peak session energy usage

$$= \frac{\sum E_2}{t} = \frac{14.173}{10} = 1.4173 \text{ kWh/day}$$

$$\text{Average off-peak session energy usage} = \frac{\sum E_3}{t} = \frac{6.128}{10} = 0.6128 \text{ kWh/day}$$

Average day session usage cost = 1.0635 × 8.99 = 9.56 LKR/day
Average peak session usage cost = 1.4173 × 11.11 = 15.75 LKR/day
Average off-peak session usage cost = 0.6128 × 8.06 = 4.94 LKR/day
Using Eqs. (4.27) and (4.28)

$$\begin{aligned}\text{Average electricity usage} = E_{avg} &= \frac{\sum E_1 + \sum E_2 + \sum E_3}{t} \\ &= 1.0635 + 1.4173 + 0.6128 = 3.0936 \text{ kWh/day}\end{aligned}$$

$$\begin{aligned}\text{Average daily cost} = C &= \frac{x_1 \sum E_1 + x_2 \sum E_2 + x_3 \sum E_3}{t} = 9.56 + 15.75 + 4.94 \\ &= 30.25 \text{ LKR}\end{aligned}$$

Using Eqs. (4.29) and (4.30)
Cost prediction for the month$(P) = k_1(MD) + k_2 + nC$
n = 27 (since the bill is generated on midnight 27 day of the month)
MD = 0 (since domestic consumers)

$$P = 150 + 27 \times 30.25 = 966.75 \text{ LKR}$$

If the consumer desired amount is 1500 LKR

$MD_m = 0$ (since domestic consumers)

$$\text{The number of remaining days} = N = \frac{A - k_1(MD_m) - k_2}{C} - t$$
$$= \frac{1500 - 150}{30.25} - 10 = 34.6\text{days}$$

References

1. ADE 7758 Data Sheet, www.analog.com/static/imported-files/data_sheets/ADE7758.pdf. Accessed 03 July 2012
2. V180MA3B Data Sheet, doc.chipfind.ru/pdf/littelfuse/v180ma3b.pdf. Accessed 10 July 2012
3. IEEE Std 1459-2000 (2000) IEEE Trial-Use Standard Definitions for the Measurement of Electric Power Quantities Under Sinusoidal, Nonsinusoidal, Balanced, or Unbalanced Conditions. kazus.ru/nuke/modules/Downloads/pub/148/0/IEEE%201459.PDF. Accessed 10 July 2012
4. Emanuel AE (1993) Apparent and reactive powers in three-phase systems. In: Electrical power, Euro. Trans. Electr. Power 3:7–14
5. Microchip 18F452 data Sheet, http://ww1.microchip.com/downloads/en/devicedoc/39564c.pdf. Accessed 03 Sept 2011
6. PCF 8583 clock calendar IC data sheet, www.nxp.com/documents/data_sheet/PCF8583.pdf. Accessed 06 Sept 2011
7. Sim 900 reference Design Guide, wm.sim.com/Sim/News/photo/2010222155735.pdf. Accessed 08 Sept 2011

Chapter 5
Short-Term Electricity Demand Forecasting and Warning Signal Generation

Abstract Short-Term Electricity Demand Forecasting (STEDF) provides many advantages for electricity suppliers as well as consumers. In this chapter the authors focus on STEDF and how it contributes to demand side load management. Different types of electricity demand forecasting methods are highlighted. A case study is done by selecting a medium voltage industrial consumer to illustrate the applications of STEDF. The model developed for demand forecasting can be used with smart meters to forecast the demand, calculate the maximum demand and to control the demand side loads. Furthermore the forecasted demand can also be used to generate warning signals regarding maximum demand. Ultimately the proposed system will help reduce the demand and save the electricity bill for both residential and industrial consumers.

5.1 Introduction

Industrial consumers are looking for better energy management practices to reduce the cost of electricity. The electricity bill is prepared considering both active energy and maximum demand by some power utilities. Maximum demand is like a penalty for industrial consumers where they have poor power factor. They get extra reactive power from the system to drive their machinery apart from real power. This will cause additional current flow in the supply cables and increase heat loss. With the improved power factor, the maximum demand can be reduced to some extent. It could be the installment of a capacitor bank at the consumer end. However this kind of mechanism provides only the reactive energy required. Although some industries have good power factors (above 0.9), they face problems due to unexpected maximum demand which is higher than their average monthly demand. Due to simultaneous start of heavy electrical machinery, failure in already installed capacitor banks and run of non-essential loads at peak demand

K. S. K. Weranga et al., *Smart Metering Design and Applications*,
SpringerBriefs in Applied Sciences and Technology,
DOI: 10.1007/978-981-4451-82-6_5,

time will lead higher maximum demands. By using a system to measure the abnormal change in demand with STEDF methods, the maximum demand charge can be reduced under above conditions. It will be an additional benefit if the system can warn consumer before the end of demand calculation time period where there is a potential of rise in demand.

When it comes to the residential electricity consumers, the STEDF will be beneficial where they can save on electricity cost under real time price schemes. For an example, activation of an alarm signal when there is a potential of rise in demand over the user defined demand and the tariff is over the specified level, will warn the consumer. This kind of mechanism will help switch off non-essential loads in peak hours thereby reduce their monthly bills. On the other hand STEDF methods contribute smart meters to take control actions when meters have access to appliance and equipment controllers. It can be done by reducing loads to pre-set levels, switching off non-essential loads, cycling loads on and off according to pre-set timing schedules and sending waiting signals to loads at higher demands and higher tariff rates.

5.2 Electricity Demand Forecasting Methods

A wide variety of electricity forecasting methods can be found with different complexity and estimation procedures. The accuracy may vary one method to another where it is applied. In forecasting as a discipline 6 factors should be satisfied. These factors are used to measure the forecast performance. They are objectivity, validity, reliability, accuracy, confidence, and sensitivity. The objectivity means that the results of a forecasting entirely depend on the data rather on the person who conduct the forecast. Validity determines whether the forecast approximates the series that is interested. Reliability is measured by the consistency of the results. Accuracy is used to measure the fitness of the forecast. Confidence tells what probability should be accepted the results. Sensitivity is highly related to the method than the results [1]. Following are some of forecasting methods which have also been used in electricity demand forecasting

- Multiple regression
- Exponential smoothing
- Stochastic time series
- Fuzzy logic
- Neural networks
- Knowledge-based expert systems.

(A) Multiple regression

Weighted least squares estimation is used to develop a statistical relation between the load and the weather conditions and the day type influences in multiple regression analysis. The regression coefficients are calculated using the defined amount of historical data [2]. Mbamalu and El-Hawary used (5.1) to model the load forecasting in 1993 [3].

$$Y_t = v_t a_t + \in_t \tag{5.1}$$

where;

t is the sampling time
Y_t is the measured system total load
v_t is the vector of adapted variables (time, temperature, light intensity etc…)
a_t is the transposed vector of regression coefficients
$\in_t$ is the modeling error at t

(B) Exponential smoothing

This technique can be applied to time series data. Even though this method is commonly applied to economic data, this can also be applied in electricity demand forecasting. In this model recent observations are given more weight in forecasting than older ones. This method has three different forecasting models which are single exponential smoothing, double exponential smoothing, and triple exponential smoothing. The simplest form of the exponential smoothing (single exponential smoothing) can be given by [4]

$$S_t = \alpha x_t + (1 - \alpha)S_{t-1} \tag{5.2}$$

where;

S_t is the forecast for next time period
x_t is actual value
t is the current time period
α is the smoothing factor, $0 < \alpha < 1$

(C) Neural networks

Neural networks (NNs) or artificial neural networks (ANNs) are composed of artificial neurons or nodes. There are multiple hidden layers in the networks which have many neurons [2]. ANNs are data driven and self adaptive. They can be used when the relation between data is hard to describe because NNs have the capability to learn from examples and capture the subtle functional relationships. NNs are used in electricity load forecasting since they have powerful pattern classification, pattern recognition capabilities and satisfactory performance in forecasting [5].

(D) Fuzzy logic

Several methods have been proposed for load forecasting using fuzzy logic [6]. Fuzzy logic systems have great capability in finding similarities for bulk data. They are commonly used when the mathematical model is nonlinear or poorly understood. There are two stages in fuzzy logic forecasting, which are training and on-line forecasting. The output pattern will be generated when the most probably matching pattern is found [2].

(E) Stochastic time series

The time series approach has been used over the years for forecasting in many areas including electricity demand due to easiness in understanding, implementation and accuracy of its results [7]. There are several methods used for modeling the time series which are autoregressive (AR) model, moving average (MA) model, autoregressive moving average (ARMA) model, and autoregressive integrated moving average (ARIMA) model.

(i) Autoregressive model

This model is applied when the load is assumed to be a linear combination of previous loads [2]. AR model can be written as [8]

$$X_t = c + \sum_{i=1}^{p} \varphi_i X_{t-i} + \varepsilon_t \tag{5.3}$$

where;

X_t is the predicted load at time t

$\varphi_i, i = 1, \ldots, q$ are model parameters to be found

c is a constant

ε_t is the random variable also known as white noise

(ii) Moving average model

MA model is conceptually linear regression of current value against current and previous value white noise. These white noise or error terms are considered to be random which typically follow the normal distribution. The moving-average of order q can be written as [9]

$$X_t = \mu + \varepsilon_t + \sum_{i=1}^{q} \theta_i \in_{t-i} \tag{5.4}$$

where;

X_t is the predicted load at time t

$\theta_i, i = 1, \ldots, q$ are model parameters to be found

μ is the expectation of X_t

$\in_t, \in_{t-1}, \ldots$ are error terms

(iii) Autoregressive moving average model

ARMA model is widely used for forecasting of economic and other time series. The ARMA (p, q) model with p autoregressive terms and q moving average terms can be written as [10]

$$X_t = c + \varepsilon_t + \sum_{i=1}^{p} \varphi_i X_{t-i} + \sum_{i=1}^{q} \theta_i \in_{t-i} \tag{5.5}$$

All the terms have typical meaning as described in (5.3) and (5.4). The parameters of this model are identified using the recursive scheme, or using a maximum-likelihood approach [2].

(iv) Autoregressive integrated moving average model
ARIMA model is a generalization of ARMA model where it could be applied even for non stationary time series. The general form of this model is ARIMA (p, d, q) where p, d, and q are parameters of order of autoregressive, integrated, and moving average terms respectively [11]. The ARIMA (p, d, q) model is written as [12]

$$\emptyset(B)(w_t - \mu) = \theta(B)a(t) \tag{5.6}$$

where;

t is the time index
B is the backshift operator
w_t is the response series after differencing
$\emptyset(B)$ is the autoregressive operator
$\theta(B)$ is the moving average operator
$a(t)$ are the sequence of random errors

(F) Knowledge-based expert systems
This kind of a system is basically a computer program which has the ability to reason, explain, and knowledge on information available to it. It is built by a knowledge engineer who extracts load forecasting knowledge from an expert in the field. This knowledge is represented as facts and rules. The rules are named as "IF–THEN" rules. Some of the rules change over the time while others don't [2].

5.3 A Case Study on How the STDF is Applied to Reduce the Maximum Demand Charge

(A) Data preparation
Historical data of electricity demand should be prepared in order to select a proper model for forecasting. The data can be recorded in minute basis or less over long time periods to have a good accuracy on forecasting. This could be the active power, active energy or apparent power, apparent energy of a particular consumer. After the collection of the data, a plot on demand against the time can be prepared to visualize clear patterns in the demand. For an example some consumers typically follow the same load pattern for each week day. The load behaviors in weekend also have closer relationships. In most cases we can go for a simple method than a complex one by identifying the regular patterns in data. Otherwise when there are no significant patterns, we can follow complex forecasting methods like neural networks, fuzzy logics etc....

(B) Demand analysis for a porcelain industry

Medium voltage (33 kV) porcelain industry was selected for the data measurements. The major electricity consuming units are the 'Biscuit kiln' and the 'Decoration kiln'. Other dominant consumers are nine units of ball mills and filter units including preparation machines. According to the consumption data for 2009–2011, the average monthly electricity consumption was around 588 MWh and the average monthly maximum demand was 1514 kVA. Remote metering records were taken to analyze the demand profile for 6 different days starting form 19.04.2011. These remote energy measurements were taken at 33 kV side of the transformers. The consumption and demand data were available at 15 min intervals. The load profile of the factory is shown in Fig 5.1.

We can clearly see that the demand pattern of this industry is the same for all week days. The demand in Saturday is much similar to other week days up to 1300 h. This is because the plant is closed for Saturday evenings. The demand in Sunday is the lowest among the other days because it is also a non-working day for this factory.

Electric kilns are run through out the day at around 550 kVA. Therefore, nearly 42 % of the total electricity demand is due to electric kilns. The factory operates ball mills in its preparation department which can be considered as a continuous load in the factory when the factory is in its full flow. When the preparation department is operating at full load, it adds extra 150 kVA to the factory load. The typical operating period of the crushers is between 3.30 pm and 7.30 pm. The demand of crushers is around 70 kVA. Other loads are rollers, hydraulic pressures and air compressors. Therefore the average apparent power of the factory during normal operating hours is around 1300 kVA.

Furthermore, it can be noticed that there is a sharp rise in the electricity demand with the start of the work in the morning. The variation of demand of the factory can be more clearly visualized when a single normal working day is considered. The apparent power demand recorded at start of the factory is around 1350 kVA.

Considering the weekday load pattern we can figure it out four different peak sessions occur from 7.30 a.m to 4.30 p.m as shown in the Table 5.1.

The demand variation in each session can be clearly visualized by taking the average demand and plot against the time. Therefore we have selected 4 days (Wednesday, Thursday, Friday and Monday) and taken the average demand in each time slot. Then we have found the best fit curve for each session. These curves show a clear relationship between the time and the demand. It can be found

Table 5.1 Sessions of peak demands

Session	Period	Minimum load (kVA)	Maximum load (kVA)
1 (start–tea break)	7.30–9.45 a.m	1125.605	1318.174
2 (tea break–lunch)	10.00 a.m–12.00 p.m	1238.09	1291.09
3 (lunch–tea break)	1.00–3.00 p.m	992.95	1251.89
4 (tea break–end)	3.15–4.30 p.m	1052.25	1291.9

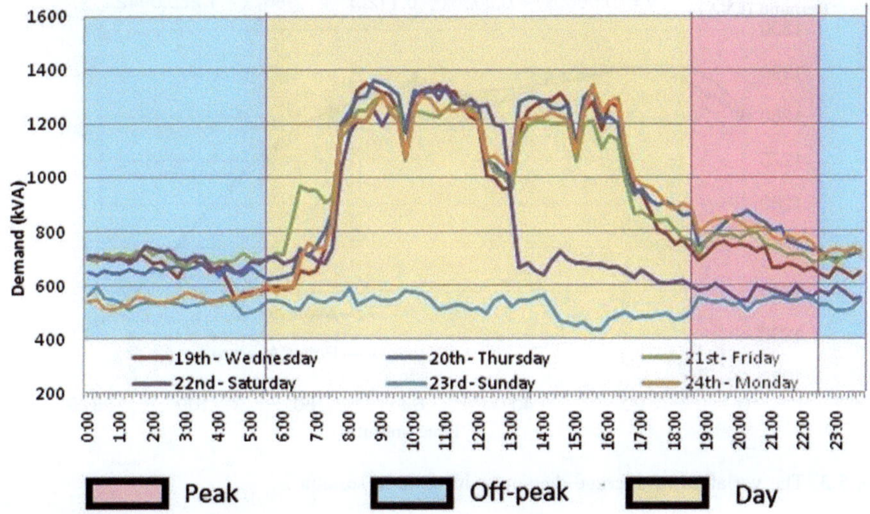

Fig. 5.1 The load profile of the factory with time of use tariff scheme

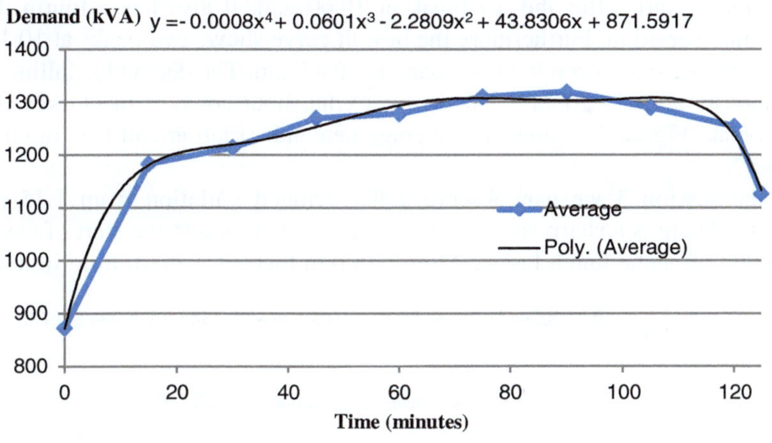

Fig. 5.2 The variation of average demand with time in session 1

that the demand variation against the time is a fourth degree polynomial for each session. However the coefficients can vary from session to session. Figures 5.2, 5.3, 5.4 and 5.5 show the demand variation and best fit curve for each session.

Fist peak occurs due to the start of the factory machinery in the morning session. It starts at 7.30 a.m and ends at 9.45 a.m. The rollers, hydraulic pressures and air compressors cause sharp rise in demand at the start apart from the base load of kilns. This sharp increase in demand can be seen in the Fig. 5.1 between 7.30 and 7.45 a.m in weekdays and Saturday. Then the demand rises to the maximum

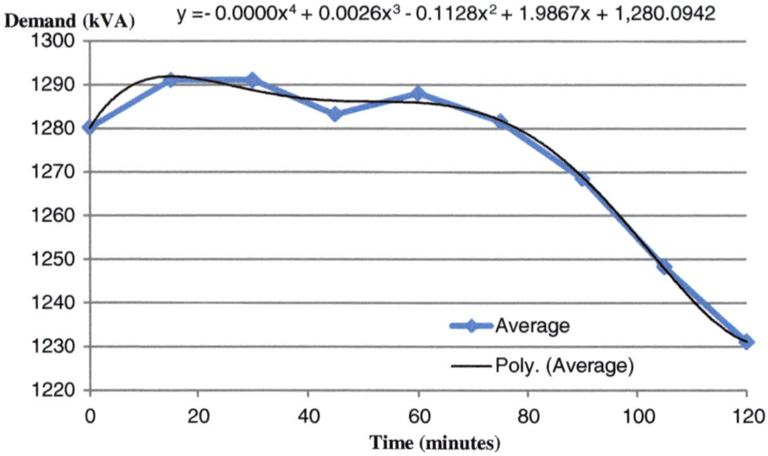

Fig. 5.3 The variation of average demand with time in session 2

value 1318.174 kVA at 9.00 a.m. The decrease in demand after 9.30 a.m is due to the shutdown of rollers and hydraulic pressures. The average demand variation and the best fit curve for session 1 are shown in Fig. 5.2.

Session 2 starts after the tea break at 10.00 a.m. It also has a fourth degree polynomial variation. Furthermore the best fit curve shows two peaks at 10.15 and 11.15 a.m. There is a drop in the demand at 10.45 a.m. The demand is falling down continuously after 11.30 a.m. This is due to the shout down of machinery for the lunch break. Figure 5.3 shows the average demand variation and best fit curve of session 2.

In the session 3 we can observe a flat demand variation from 1.15 pm to 2.45 p.m. There is a sharp rise in demand at 1.00 p.m due to the start of electrical machinery after the lunch break. After 2.45 p.m there is a sharp drop in demand.

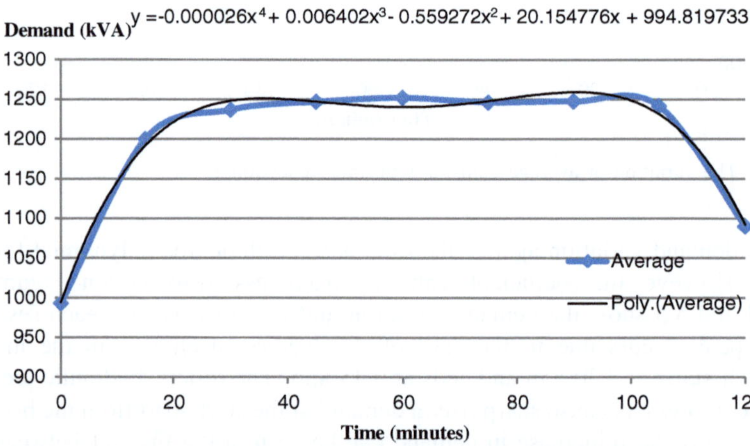

Fig. 5.4 The variation of average demand with time in session 3

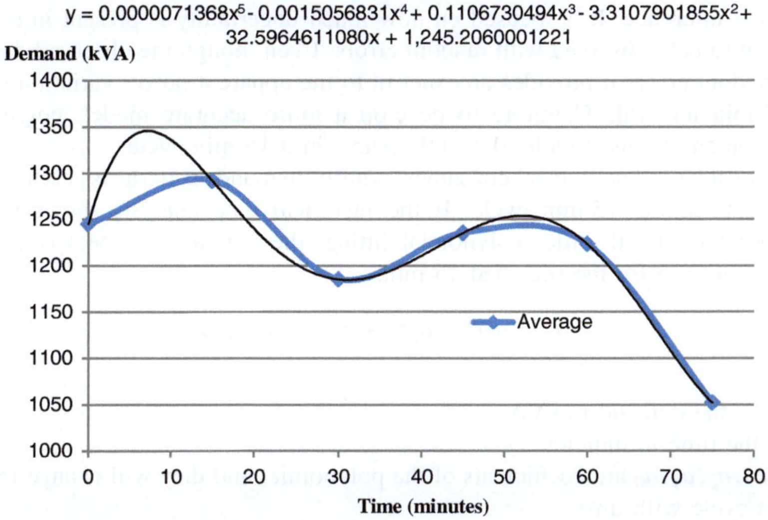

Fig. 5.5 The variation of average demand with time in session 4

This is due to the shutdown of machinery for the evening tea break. The demand variation and the best fit curve are shown in Fig. 5.4. The best fit curve of the average demand is a fourth degree polynomial.

The average demand variation and best fit curve of the session 4 are shown in Fig. 5.5. In this session, fitting a 5th degree polynomial for average demand gives more accuracy than 4th degree polynomial. The peak occurs at 3.20 p.m and its magnitude is about 1340 kVA. This is due to the start of ball mills or compressors or hydraulic pressures. There is a small drop in demand at 3.45 p.m.

(C) Mathematical model with polynomial regression

(i) Model identification

A mathematical model is used to come up with the best possible forecast which the smart meter developers can accept with reasonable confidence. Curve fitting is achieved using the polynomial regression. This is selected because the average demand in peak sessions has a 4th order polynomial variation. However this methodology can be modified for high orders as well. It is found that fitting a 4th order polynomial gives a good accuracy compared to the lower order polynomials. When the order is greater than the 4th order, the accuracy has small effects with the order. On the other hand when the order is high, more calculation time and more sample points are required. This will be a disadvantage when this methodology is implemented for microcontrollers in smart meters. Therefore 4th order polynomial fitting is selected considering the above factors.

(ii) Polynomial regression

In polynomial regression, it is assumed that the trend in the past will persist into the future [13]. It can be used where an exact fit to the data is required with

statistical inference. It is focused on how much uncertainty is present in a curve that is fit to data observed with random errors. Even though the observed data are with random errors it provides an exact fit to the apparent power variation within the 15 min interval. Therefore to develop a more accurate model, polynomial regression can be used with 10 sample points in a 15 min cycle.
Now consider a particular consumer whose demand pattern is highly time dependant for the 15 min cycle. If the historical data analysis shows a good accuracy for the 4th order polynomial fitting, the demand can be written as a function of time for the specified 15 min.

$$f(t) = a_4t^4 + a_3t^3 + a_2t^2 + a_1t + a_0 \tag{5.7}$$

where;

$f(t)$ is the demand in kVA

t is the time in minutes

a_4, a_3, a_2, a_1, a_0 are coefficients of the polynomial and they will change in each 15 min cycle with time.

Now we can write 4 equations considering the average values to construct the 4th order polynomial.

$$a_4\overline{t^4} + a_3\overline{t^3} + a_2\overline{t^2} + a_1\overline{t} + a_0 = \overline{f} \tag{5.8}$$

$$a_4\overline{t^5} + a_3\overline{t^4} + a_2\overline{t^3} + a_1\overline{t^2} + a_0\overline{t} = \overline{ft} \tag{5.9}$$

$$a_4\overline{t^6} + a_3\overline{t^5} + a_2\overline{t^4} + a_1\overline{t^3} + a_0\overline{t^2} = \overline{ft^2} \tag{5.10}$$

$$a_4\overline{t^7} + a_3\overline{t^6} + a_2\overline{t^5} + a_1\overline{t^4} + a_0\overline{t^3} = \overline{ft^3} \tag{5.11}$$

$$a_4\overline{t^8} + a_3\overline{t^7} + a_2\overline{t^6} + a_1\overline{t^5} + a_0\overline{t^4} = \overline{ft^4} \tag{5.12}$$

where;

$$\overline{t} = \frac{1}{n}\sum t_i \tag{5.13}$$

$$\overline{t^2} = \frac{1}{n}\sum t_i^2 \tag{5.14}$$

$$\overline{t^3} = \frac{1}{n}\sum t_i^3 \tag{5.15}$$

$$\overline{t^4} = \frac{1}{n}\sum t_i^4 \tag{5.16}$$

$$\overline{t^5} = \frac{1}{n}\sum t_i^5 \tag{5.17}$$

$$\overline{t^6} = \frac{1}{n}\sum t_i^6 \tag{5.18}$$

$$\overline{t^7} = \frac{1}{n}\sum t_i^7 \tag{5.19}$$

$$\overline{t^8} = \frac{1}{n}\sum t_i^8 \tag{5.20}$$

$$\bar{f} = \frac{1}{n}\sum f_i \tag{5.21}$$

$$\overline{ft} = \frac{1}{n}\sum t_i f_i \tag{5.22}$$

$$\overline{t^2 f} = \frac{1}{n}\sum t_i^2 f_i \tag{5.23}$$

$$\overline{t^3 f} = \frac{1}{n}\sum t_i^3 f_i \tag{5.24}$$

$$\overline{t^4 f} = \frac{1}{n}\sum t_i^4 f_i \tag{5.25}$$

n is referred to number of sample points
These equations can be converted to a matrix form as follows:

$$\begin{bmatrix} \overline{t^4} & \overline{t^3} & \overline{t^2} & \bar{t} & 1 \\ \overline{t^5} & \overline{t^4} & \overline{t^3} & \overline{t^2} & \bar{t} \\ \overline{t^6} & \overline{t^5} & \overline{t^4} & \overline{t^3} & \overline{t^2} \\ \overline{t^7} & \overline{t^6} & \overline{t^5} & \overline{t^4} & \overline{t^3} \\ \overline{t^8} & \overline{t^7} & \overline{t^6} & \overline{t^5} & \overline{t^4} \end{bmatrix} \begin{bmatrix} a_4 \\ a_3 \\ a_2 \\ a_1 \\ a_0 \end{bmatrix} = \begin{bmatrix} \bar{f} \\ \overline{tf} \\ \overline{t^2 f} \\ \overline{t^3 f} \\ \overline{t^4 f} \end{bmatrix}$$

Considering one particular 15 min cycle, 10 sample points can be taken from $t = 1$ to $t = 10$ min to fit the curve. Thereafter the demand and the area are forecasted for the next 5 min. The sampling and forecasting method is illustrated in Fig. 5.6.
Here, $n = 10$, and the matrix reduces to:

$$\begin{bmatrix} 2533.3 & 302.5 & 3.85 & 5.5 & 1 \\ 22082.5 & 2533.3 & 302.5 & 3.85 & 5.5 \\ 197840.5 & 22082.5 & 2533.3 & 302.5 & 3.85 \\ 1808042.5 & 197840.5 & 22082.5 & 2533.3 & 302.5 \\ 16773133 & 1808042.5 & 197840.5 & 22082.5 & 2533.3 \end{bmatrix} \begin{bmatrix} a_4 \\ a_3 \\ a_2 \\ a_1 \\ a_0 \end{bmatrix} = \begin{bmatrix} \bar{f} \\ \overline{tf} \\ \overline{t^2 f} \\ \overline{t^3 f} \\ \overline{t^4 f} \end{bmatrix}$$

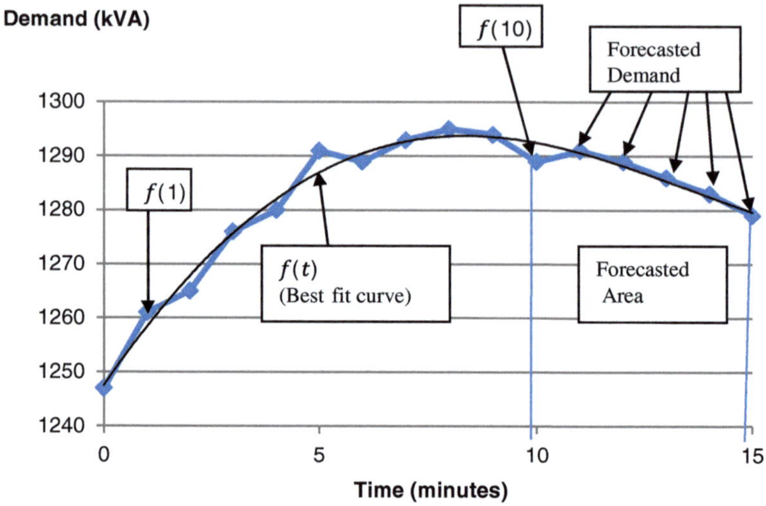

Fig. 5.6 Data sampling using demand pattern

Taking the inverse of the matrix

$$\begin{bmatrix} a_4 \\ a_3 \\ a_2 \\ a_1 \\ a_0 \end{bmatrix} = \begin{bmatrix} 2533.3 & 302.5 & 3.85 & 5.5 & 1 \\ 22082.5 & 2533.3 & 302.5 & 3.85 & 5.5 \\ 197840.5 & 22082.5 & 2533.3 & 302.5 & 3.85 \\ 1808042.5 & 197840.5 & 22082.5 & 2533.3 & 302.5 \\ 16773133 & 1808042.5 & 197840.5 & 22082.5 & 2533.3 \end{bmatrix}^{-1} \begin{bmatrix} \overline{f} \\ \overline{tf} \\ \overline{t^2 f} \\ \overline{t^3 f} \\ \overline{t^4 f} \end{bmatrix}$$

where;

$$\bar{f} = \frac{1}{10}[f(1) + f(2) + f(3) + f(4) + f(5) + f(6) + f(7) + f(8) + f(9) + f(10)] \tag{5.26}$$

$$\overline{ft} = \frac{1}{10}[f(1) + 2f(2) + 3f(3) + 4f(4) + 5f(5) + 6f(6) + 7f(7) + 8f(8) + 9f(9) + 10f(10)] \tag{5.27}$$

$$\overline{ft^2} = \frac{1}{10}\begin{bmatrix} f(1) + 2^2 f(2) + 3^2 f(3) + 4^2 f(4) + 5^2 f(5) \\ +6^2 f(6) + 7^2 f(7) + 8^2 f(8) + 9^2 f(9) + 10^2 f(10) \end{bmatrix} \tag{5.28}$$

$$\overline{ft^3} = \frac{1}{10}\begin{bmatrix} f(1) + 2^3 f(2) + 3^3 f(3) + 4^3 f(4) + 5^3 f(5) \\ +6^3 f(6) + 7^3 f(7) + 8^3 f(8) + 9^3 f(9) + 10^3 f(10) \end{bmatrix} \tag{5.29}$$

$$\overline{ft^4} = \frac{1}{10}\left[\begin{array}{c} f(1) + 2^4 f(2) + 3^4 f(3) + 4^4 f(4) + 5^4 f(5) \\ +6^4 f(6) + 7^4 f(7) + 8^4 f(8) + 9^4 f(9) + 10^4 f(10) \end{array}\right] \tag{5.30}$$

The coefficients become

$$\begin{aligned} a_4 = {} & 0.003f(1) - 0.003f(2) - 0.002f(3) + 0.000f(4) + 0.003f(5) \\ & + 0.003f(6) + 0.000f(7) - 0.002f(8) - 0.003f(9) + 0.003f(10) \end{aligned} \tag{5.31}$$

$$\begin{aligned} a_3 = {} & -0.006f(1) + 0.073f(2) + 0.061f(3) - 0.004f(4) - 0.005f(5) \\ & - 0.060f(6) - 0.016f(7) + 0.048f(8) + 0.068f(9) - 0.049f(10) \end{aligned} \tag{5.32}$$

$$\begin{aligned} a_2 = {} & 0.579f(1) - 0.553f(2) - 0.525f(3) - 0.040f(4) + 0.369f(5) \\ & + 0.446f(6) + 0.158f(7) - 0.290f(8) - 0.464f(9) + 0.310f(10) \end{aligned} \tag{5.33}$$

$$\begin{aligned} a_1 = {} & -2.079f(1) + 1.491f(2) + 1.619f(3) + 0.373f(4) - 0.816f(5) \\ & - 1.159f(6) - 0.507f(7) + 0.644f(8) + 1.162f(9) - 0.728f(10) \end{aligned} \tag{5.34}$$

$$\begin{aligned} a_0 = {} & 2.499f(1) - 0.833f(2) - 1.249f(3) - 0.417f(4) + 0.499f(5) \\ & + 0.833f(6) + 0.417f(7) - 0.417f(8) - 0.833f(9) + 0.499f(10) \end{aligned} \tag{5.35}$$

The equation for the polynomial can be found if the values are known from $f(1)$ to $f(10)$. However to generate a warning signal regarding the maximum demand, the average apparent energy within 15 min should be calculated. Therefore it is necessary to forecast the apparent energy from $t = 10$ min to $t = 15$ min. To find the apparent energy for the next 5 min when the time equals 10 min, the polynomial should be integrated form $t = 10$ to $t = 15$ min.
If the forecasted apparent energy in next 5 min is A_1

$$A_1 = a_4 \int_{10}^{15} t^4 dt + a_3 \int_{10}^{15} t^3 dt + a_2 \int_{10}^{15} t^2 dt + a_1 \int_{10}^{15} t^1 dt + a_0 \int_{10}^{15} t^0 dt \tag{5.36}$$

By replacing coefficients

$$\begin{aligned} A_1 = {} & 18.382f(1) - 27.908f(2) - 16.755f(3) + 10.406f(4) + 26.872f(5) \\ & +20.672f(6) - 5.435f(7) - 33.956f(8) - 32.667f(9) + 45.387f(10) \end{aligned} \tag{5.37}$$

The units of A_1 is kVA min
Digital meters have different sampling rates and discrete time equations for calculating the apparent energy and the accuracy may differ in different meters accordingly. Hence, (5.38) is used for better accuracy considering the continuous demand signal.

$$A_0 = \int_{t=0}^{t=10} F(t)dt \tag{5.38}$$

A_0 *is* referred to apparent energy over 10 min in kVA min and $F(t)$ is referred to apparent power signal.
Now the forecasted average demand (D_f) can be written as

$$D_f = (A_0 + A_1)/15 \tag{5.39}$$

Ultimately (5.31)–(5.35) can be used to derive the forecasted demand equation. For the maximum demand warning signal, (5.39) can be used.

(D) Algorithm of STDF for smart meters
This study focuses on development of an algorithm that can be easily implemented on smart meter software/hardware system. The input parameter for the system is electricity demand. The developers can use either smart meter hardware or data obtained via a communication link like RS232. Therefore with smart meters we can easily get the apparent power samples in regular time intervals to construct the forecast demand equation. The GSM network can be used to transfer the warning signal to the consumer via SMS. However there are certain questionable issues regarding GSM network such as its scalability, reliability and security, especially under high load [14].

Smart meters record energy consumption and power quality at a preset interval, usually an hour or less. The characterization of patterns useful for consumption analysis and demand forecasting is initiated by the collection of such meter readings [15]. We can use the this kind of data to read values from $f(1)$ to $f(10)$. If we consider 1 h period, there are four 15 min intervals, each starting at $t = 0$, $t = 15$, $t = 30$ and $t = 45$ min. The RTC source may be used to generate interrupts on minute basis. At the beginning of each 15 min, the calculation should be initiated. Therefore $f(1)$ is read at 1 or 16 or 31 or 46 min. $f(2)$ is read at 2 or 17 or 32 or 47 min. Identical patterns can be followed to read values from $f(3)$ to $f(10)$. Ultimately when the readings are taken they can be used for polynomial fitting.

When $t = 10$, 25, 40 or 55 min, the A_0 and A_1 can be found. D_{u1} is the user defined demand level. If $D_{u1} < D_f$ condition is satisfied the maximum demand warning signal is generated. The next condition is to check whether the demand and current tariff exceed the user defined level. As we know some power suppliers use time of use tariff schemes. The tariff may vary on hourly basis or on minute basis. However the suggested system should check the tariff in each of the 15 min interval. This is because the smart meters are normally fed with hourly tariff or daily tariff ahead. If the tariff $(C(t))$ and the demand $(f(t))$ are going to exceed the user defined demand value (D_{u2}) and user defined tariff level (C_u), the smart meter can warn the consumer or control the demand side loads to save electricity costs. Otherwise it follows the next instruction of the program. The next instructions may be the other part of the smart meter main software. This algorithm can be used with an interrupt routine when using a microcontroller. The algorithm is shown in Fig. 5.7.

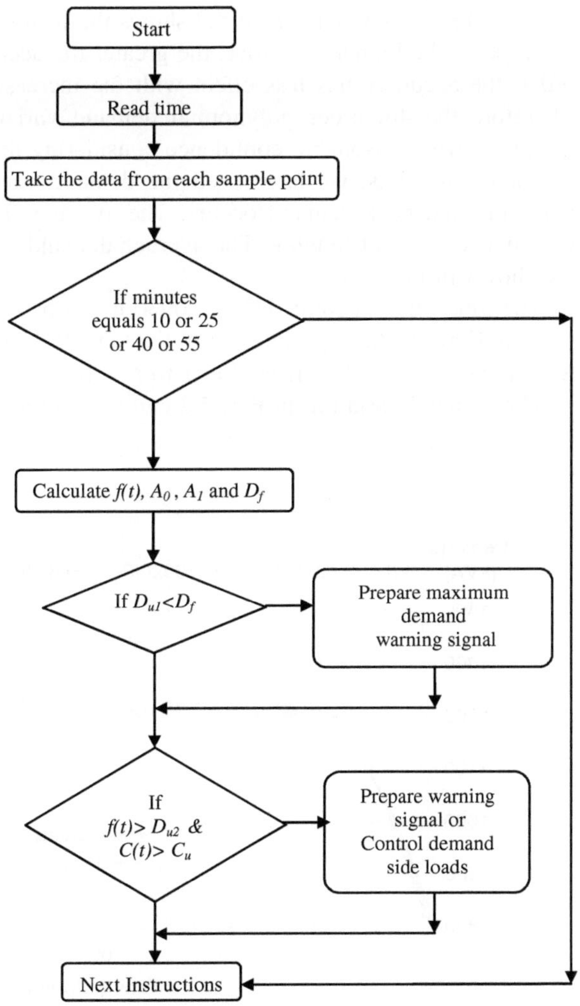

Fig. 5.7 Demand forecasting algorithm

Table 5.2 Probabilistic results

Order of the polynomial	Probability of 95 % accuracy
1	0.4128
2	0.8173
3	0.8934
4	0.9563
5	0.9564
6	0.9564

(E) Results and discussions

Percentage values of accuracy are found for each case considering the order of the polynomial. The probability of getting 95 % accuracy is calculated against the order of the polynomial. Table 5.2 shows the results obtained.

As seen, the higher the order, the greater the accuracy. However, after the 4th order, the accuracy has less effect with the increasing order of the polynomial. Therefore the 4th order polynomial demand variation can be applied for this consumer with reasonable confidence considering the data for 6 days.

For calculations, we have used the demand variation from 0730 to 0930 h, when the first peak demand occurs. The average values are taken for the above time interval except Sunday. The average demand variation and the best fit curve are shown in Fig. 5.8.

The time interval from 0730 to 0930 h can be divided into 8 sub intervals of 15 min. Here the first sub interval (0730 to 0745 h) is selected to find $f(t)$ and D_f. Sample points are taken from $t = 1$ to $t = 10$.

The demand variation in Fig. 5.8 can be written as

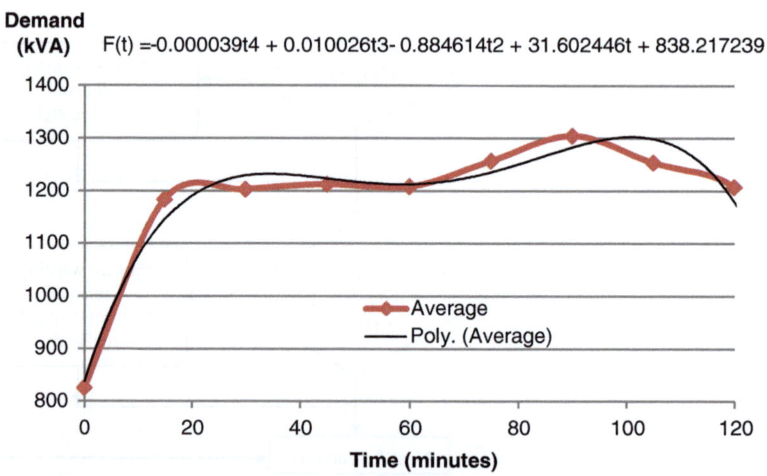

Fig. 5.8 The variation of average demand with time

$$F(t) = -0.000039t^4 + 0.010026t^3 - 0.884614t^2 + 31.602446t + 838.2172 \tag{5.40}$$

Using (5.31)–(5.35) we can calculate the coefficients.

$$a_4 = -0.000040,\ a_3 = 0.010057,\ a_2 = -0.878208,\ a_1 = 31.507987,$$
$$a_0 = 838.114272$$

Using (5.7), the forecasted demand equation can be written.

$$f(t) = -0.000040t^4 + 0.010057t^3 - 0.878208t^2 + 31.507987t + 838.114272 \tag{5.41}$$

Using (5.37), the forecasted area can be calculated.

$$A_1 = 5562.4890\,\text{kVAmin} \tag{5.42}$$

Using (5.38) and (5.40), the apparent energy within 10 min can be found.

$$A_0 = \int_0^{10} F(t)dt = 9691.70796\,\text{kVAmin} \tag{5.43}$$

Using (5.39), the forecasted demand can be found.

$$D_f = 1016.9464 \text{ kVA} \tag{5.44}$$

If the user defined maximum demand is below 1016.9464 kVA, the demand warning signal is generated after 10 min. Same procedure can be used to calculate the $f(t)$ and D_f for other 15 min intervals form 0745 to 0930 h.

Simulation results were obtained using MATLAB software for the demand data available from 0730 to 0930 h. Using demand variation on Monday 24th April 2011 for this period of time, the MATLAB program was written to generate the forecast demand and warning signals. When the forecasted demand exceeds the user defined demand (D_{u1}), the warning signal is displayed on the figure at the forecasted time. The actual demand variation is shown in black dots. Sample points are taken at every minute and are shown in green dots. The cross marks are the forecast demand values for the next 5 min. The results are shown in Fig 5.9. For the illustration of warning, we have defined $D_{u1} = 1000$ KVA which is always below the actual demand within the selected time period. However in the actual scenario, D_{u1} should be greater than or equal to the average maximum demand of the year.

The forecast and actual demand values can be obtained when the program executes the MATLAB code. Table 5.3 shows the data obtained using the MATLAB program.

Simulation results indicate that the accuracy of prediction results is 99.91 %. During the same time it shows an inaccuracy of 5.09 % for the given data set. However this system should be tested for more historical and more future data to

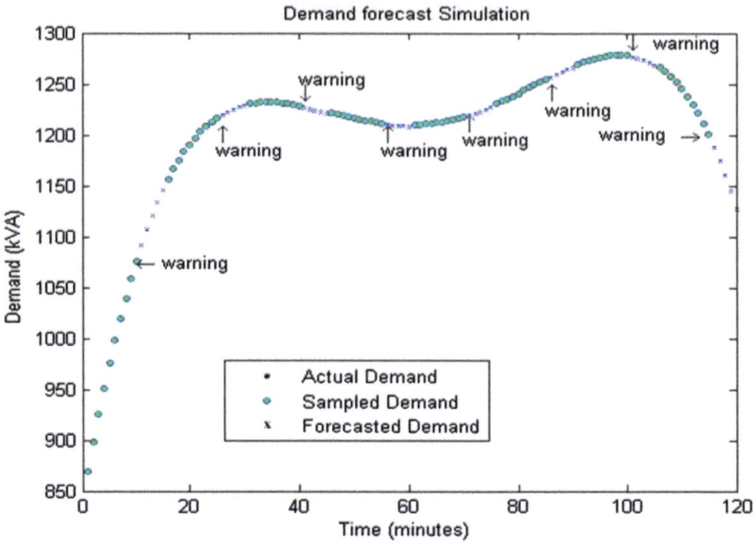

Fig. 5.9 Output figure when $D_{u1} = 1000$ KVA

Table 5.3 Simulation results

Time(minutes)	Actual average demand (kVA)	Forecasted average demand = D_f(KVA)	Accuracy (%)
0–15	1048.932	1016.9526	96.95
15–30	1144.314	1199.4992	95.18
30–45	1227.454	1229.2780	99.85
45–60	1235.679	1214.5453	98.29
60–75	1281.353	1216.1727	94.91
75–90	1321.802	1247.6467	94.39
90–105	1273.934	1275.0688	99.91
105–120	1247.920	1217.1555	97.53

Table 5.4 Time of use tariff

Day (0530–1830 h)	Peak (1830–2230 h)	Off-Peak (2230–0530 h)
10.25 LKR	13.40 LKR	7.15 LKR

get a reasonable number for accuracy. The accuracy range obtained here is sufficient for an application like warning signal generation.

In the next step of simulation, the time of day tariff scheme and user defined demand level are used to generate the warning signal. Data pattern in Fig 5.9 has been used for this simulation. However now there are two parameters to generate the warring signal which are D_{u2} and C_u.

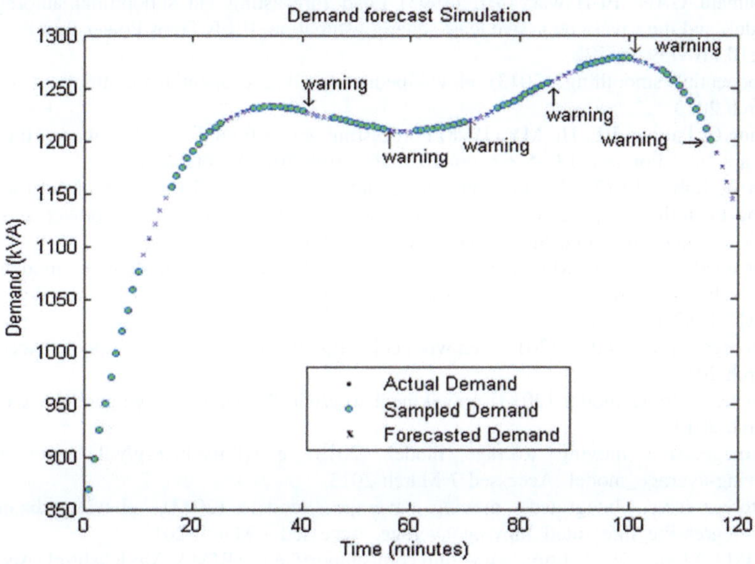

Fig.. 5.10 Output figure when $D_{u2} = 1200\,\text{kVA}$ and $C_u = 10.00\,\text{LKR}$

The time of use tariff scheme is shown in Table 5.4.

For the simulation we defined $D_{u2} = 1200$ kVA and $C_u = 10.00$ LKR. The simulation results obtained under above conditions are shown in Fig. 5.10.

A new approach for 5 min electricity demand forecasting is presented using the polynomial regression. This study will be helpful to industries that suffer from higher maximum demand charges and residential consumers who are interested in energy saving. Due to the simplicity of the methodology and reduced calculation time, it can be easily developed using smart meter software/hardware. By adding this as a feature to smart metering, warning signals can be sent to the industrial consumers via SMS or other communication link. Residential consumers benefit from the alarm signals on higher demands during peak hours. When the smart meters have access to the consumer's loads, control actions can be taken to reduce the demand. Ultimately the suggested system will help the demand side load management and electricity bill reduction. The results were obtained only by simulation. This system should be implemented and tested for long run to verify the accuracy.

References

1. Cooray TMJA (2008) Scope of the book. In: Applied time series: analysis and forecasting, Narosa Publishing House, New Delhi, India, p 1–21
2. Alfares HK, Nazeeruddin M (2002) Electric load forecasting: a literature survey and classification of methods. Int J Syst Sci 33:23–34. doi:10.1080/00207720110067421

3. Mbamalu GAN, El-Hawary ME (1993) Load forecasting via suboptimal autoregressive models and iteratively recursive least squares estimation. IEEE Trans Power Syst 8:343–348. doi:10.1109/59.221222
4. Exponential smoothing (2013) en.wikipedia.org/wiki/Exponentialsmoothing. Accessed 7 March 2013
5. Zhang G, Patuwo BE, Hu MY (1998) Forecasting with artificial neural networks: the state of the art. Int J Forecast 14:35–62. doi:10.1016/S0169-2070(97)00044-7
6. Jain A, Babita M (2012) Fuzzy modeling and similarity based short term load forecasting using evolutionary particle swarm optimization. In: Proceedings IEEE power and energy society general meeting, San Diego, CA, 22–26 July 2012
7. Mohamed A, Abu-El-Magd, Naresh KS (1982) Short-term load demand modeling and forecasting: a review. IEEE Trans Syst Man Cybern 12:370–382. doi:10.1109/TSMC.1982.4308827
8. Autoregressive model (2013) en.wikipedia.org/wiki/Autoregressive_model. Accessed 7 March 2013
9. Moving average model (2013) en.wikipedia.org/wiki/Moving-average_model. Accessed 7 March 2013
10. Autoregressive moving average model (2013) en.wikipedia.org/wiki/Autoregressive–moving-average_model. Accessed 7 March 2013
11. Autoregressive integrated moving average model (2013) en.wikipedia.org/wiki/Autoregressive_integrated_moving_average. Accessed 7 March 2013
12. ARIMA Model (2013) http://www.jmp.com/support/help/ARIMA_Model.shtml. Accessed 7 March 2013
13. Towill S (1974) Estimation of maximum demand on a British electricity-board system (forecast periods of 1–3 years). In: Proceedings of the Institute of Electrical Engineers, vol 121, pp. 609–615. doi:10.1049/piee.1974.0142
14. Sheikh SS, Sharma S (2011) Design and implementation of wireless automatic meter reading system. Eng Sci Technol Int J 3(3):2329–2334
15. De Silva D, Xinghuo Y, Alahakoon D, Holmes G (2011) Semi-supervised classification of characterized patterns for demand forecasting using smart electricity meters. In: Proceedings electrical machines and systems international conference, Beijing, 20–23 Aug 2011

Chapter 6
Smart Metering Applications

Abstract This chapter describes different types of applications related to smart metering in smart grid. The interference of smart meters in distributed generation is illustrated and discussed. The voltage monitor and control with the aid of smart meters and its benefits are also discussed. HVAC system performance enhancement with smart metering is highlighted. Residential load controlling through smart meters, the integration of smart appliance controllers, demand side primary frequency control, optimal energy management under time varying tariff structures, and their impacts on energy conservation and cost reduction are summarized with modern world examples. Different types of tariff schemes commonly used with smart metering such as real time pricing, time of use pricing, critical pricing, and day ahead pricing and their impact on demand management are compared and discussed.

6.1 Introduction

Smart meters are used in many applications in smart grid due to their multi tasking capability and high accuracy in measurements. They can measure not only the energy but also other electrical and non electrical quantities. The main advantage of smart meters is that they can contribute for energy conservation and efficiency. Smart meters have several applications such as net metering in DG, voltage monitoring at distributed grid, increasing the HVAC system performance, frequency monitoring and demand management at the consumer end.

Modern smart meters capture the overall system performance. Therefore they can also be used in micro grids. They have the capability to measure the bidirectional energy flow and have an advantage in net metering applications. Smart meters are installed at different locations in distributed grid to monitor the voltage profiles. They can actively contribute for voltage control and reactive power

K. S. K. Weranga et al., *Smart Metering Design and Applications*,
SpringerBriefs in Applied Sciences and Technology,
DOI: 10.1007/978-981-4451-82-6_6,

management to enhance the system stability and efficiency. Smart meters have a trend in installing at large buildings where high energy efficiency is required. In such systems, smart meters have the capability to measure the temperature at different locations and control the HVAC loads to improve the overall system performance.

On the other hand smart meters support residential load control and demand management. Consumers are updated with hourly or half hourly tariff, consumption details, and power outages. These information are very much helpful for consumers to change their behavior. The electricity consumption can be reduced at peak hours and shifted to off peak hours to minimize the electricity bill. Moreover the system load factor is also improved with demand management. Nowadays smart appliance controllers are used along with smart meters. These devices are capable of remote connection/disconnection of loads, electricity cost reduction under time varying price schemes, and control of demand side primary frequency.

6.2 Distributed Generation (DG)

Normally, electricity is generated in centralized power plants that exploit conventional non renewable fuels. With DG, more renewable energy is added to the power grid in any node of the distributed network [1]. DG is a good solution for growing demand for electricity which adds energy to the network from small energy sources. These micro generators are installed at households, factories, and offices. In most cases the customer uses all the output from DG system and any surpass is delivered to the grid. When the power requirement at the consumer side is higher than the generated, additional power can be taken from the grid. DG systems can be found typically in the range of 3–10,000 kW. They have lower impact on environment since most of the power comes from renewable sources such as wind, solar, and biomass [2]. The limitations of the existing grid system can be narrow downed with DG systems. They can be connected to the same grid as centralized power stations. Figure 6.1 shows the micro grid based DG system.

The measurement of energy generated, consumed, and delivered is an essential requirement in DG systems. Typical electromechanical meters can be used to measure this energy. However they have some drawbacks in performance and accuracy. Nowadays these electromechanical meters have hardly any use in DG systems. They are replaced with smart meters since smart meters have the capability to capture the overall system performance in a micro grid. They can measure other essential electric parameters such as power quality, energy flow, voltage dip and transient of the connected micro grids [3]. To establish a highly reliable network, these data should be monitored in real time or near time. On the other hand smart meters can be used for energy saving and optimization in building automation systems.

Smart meters play a key role in net metering in DG systems. Net metering is an electricity policy which is designed for energy measurement in DG systems. The

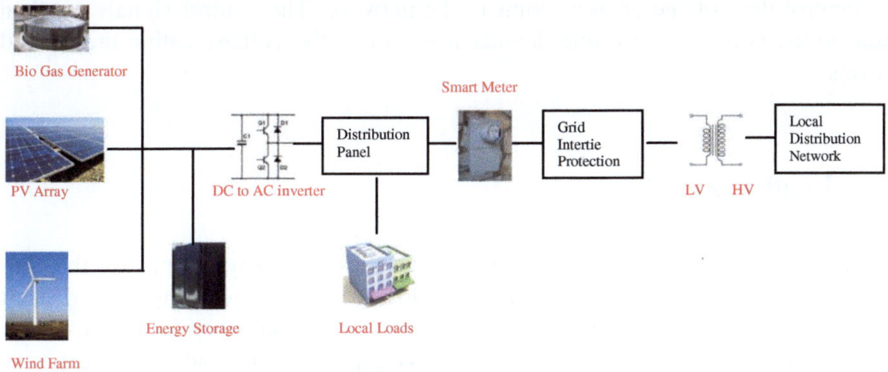

Fig. 6.1 Micro grid based DG system

term 'Net' is referred to the sum of active energy delivered and consumed. In other words this could be the deduction of any energy outflows from the metered energy inflows. A retail credit is given to the system owner for at least a portion of the electricity they generate [4]. There should be a production energy meter (optional) to measure the number of units generated by the consumer and a utility meter to measure both consumed and exported energy. The utility energy meter should be able to measure bidirectional energy flow. Nowadays due to more advanced energy measurement requirements, smart meters are used as utility meters in net metering. They can separately measure and store the consumed and exported energy.

6.3 Voltage Control in Smart Grid

The voltage control on the distribution network is a key requirement in smart grid concept. Keeping the voltage magnitude within statutory limits at feeders and consumer end fulfills the requirement of a healthy supply. On the other hand voltage control and reactive power management enhance the system stability and efficiency. This can be achieved by installing voltage control devices and power quality sensors at different locations in the grid. These devices are OLTC transformers, synchronous DG, shunt capacitors, static var compensator, converter based flexible AC transmission system controllers etc. For proper maintenance and operation, these devices require software tools and communication infrastructure [5].

In the traditional grid, voltage controllers and voltage sensors are located only at substations and sending-ends of transformers. With the use of smart meters, now the utilities are capable of monitoring the voltage and power quality measurements event at the end users. Smart meters provide real time measurements of data and communicate them to a control center. These data can be analyzed to take decision to control the voltage at each point in the network. The control signals are then sent to the voltage controlling devices to maintain the voltage within the desired levels.

6.4 Load Control

Smart metering provides much more opportunities in energy savings for the consumer as well as for the energy supplier. However smart meter, by itself doesn't save energy directly. Proper demand management is mandatory for better energy savings with information displays, time varying pricing and load control [6].

Real time information about the current energy consumption, its cost, real time price signals, estimated cost of the month, credits earned by exporting energy, and indication of any power cuts can be provided to the consumer via SMS, GPRS or any other communication link. The incoming information can be displayed at the smart meter or in-home display unit. This kind of information will help consumers to change their behavior to achieve absolute reduction in energy use or shifting energy use from peak hours to off–peak hours. Cut down of unnecessary peak in the system and the exported energy from renewable sources will lead to a reduction of carbon dioxide emission. Therefore incenting the consumer for demand management will help to achieve high efficiency, customer satisfaction and reduction of carbon dioxide.

Modern smart metering systems support information display devices, appliance and equipment controllers, and integrated load control systems. Smart appliances individually measure the power and energy consumption of each appliance at the consumer end. The data from these devices are collected at the control center in near real time. The control signals are sent back to the appliance controllers from the control center. Therefore two-way communication link should be established to communicate between the load manager and the controlled loads. Load control system can be operated by an electricity supplier or network operator, a market or system operator, a demand side response service provider or end user [6]. Special tariff rates such as real time pricing, day ahead pricing, time of use pricing, and critical peak pricing should be introduced with load control, aimed at compensating the reduced autonomy of the consumer. This allows the operator the right to control load demand [1]. Figure 6.2 shows a typical load control system combined with smart metering.

At the operating center, the load manager sends signals to initiate or terminate a load control program. The signal may also be generated automatically when the demand exceeds the pre-set load levels, excurse outside pre-set frequency or voltage parameters. Instructions are given to reduce the demand to pre-set levels, cycle the loads on and off in regular intervals, or to switch off the load completely. The communication channel might be handled by a network operator. Wired links such as internet, fiber optic, PLC or wireless links such as GSM, GPRS, RF, Wi-Fi can also be used as communication links.

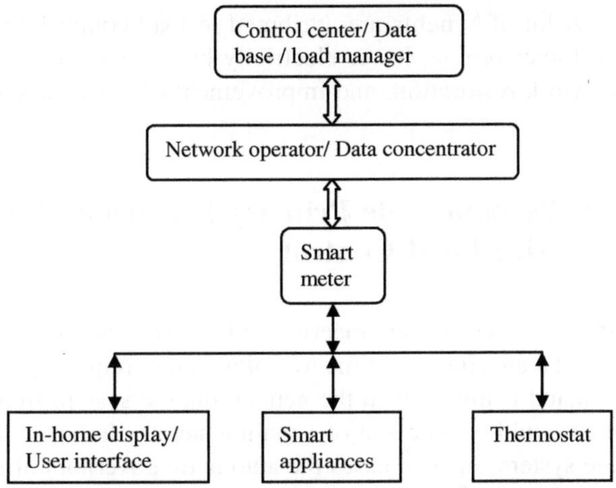

Fig. 6.2 Smart metering system with load control

In direct load control, signals are typically sent to the smart meter in which the control actions are taken. The meter selects the necessary load adjustments according to the instructions received from the load manager. However modern load control systems are designed to control the customers' appliances and equipments directly. In this scenario, coded signals are directly passed to the appliances. These smart appliances and device controllers are designed with simple on–off switches, programmable thermostats and demand controllers. The technology is being used to control appliances such as air conditioners, electric hot water systems, and pool pumps [7]. In an event of high power consumption in a particular area, the load manager can adjust the consumer loads to cut down the peak. This can be achieved by cycling off the air conditioner compressor to lower the electricity consumption for several consumers. After successful operation of a particular load management program, the acknowledgement signal is sent back to the operator. This can only be done when there is a two way communication between the consumer and the load manager [6].

The other method of load control is called as locally switched load control in which the control actions are taken by the consumer. Signals generated by the load manager are displayed on the home display or at the meter. Basically this information include hourly tariff, indication of power outages, and average power usage of the consumer. The consumer can respond to these signals by adjusting his load patterns [6]. The consumers are highly encouraged to shift the power usage to off-peak hours to minimize the electricity cost.

In DG control, more flexibility is provided with dispatching the generation and loads. Here the generation dispatch is done remotely by the operator. Optimal power flow and distribution reconfiguration systems can also be introduced. The system can be managed properly under emergency conditions and amount of curtailed load and penalties due to load shedding can be minimized [1].

A lot of benefits are attributed to load controlling including remote control of customer profile, lower electricity bill, prevention of load peaks, optimization of network restoration, and improvement of network stability.

6.5 Demand Side Primary Frequency Variation with Load Control

In power systems engineering, it is known that the demand and the generation are kept balanced in real time to maintain the frequency at its nominal value. When the demand is higher than the generation, the system frequency starts to drop rapidly. This could be an event of a main generator failure or connection of a large load to the system. In such cases the automatic governors of generators try to increase the actual power to stabilize the frequency. In addition to generators, frequency can be controlled by demand management which is known as frequency controlled demand management (FCDM). FCDM can switch off the consumer loads to stabilize the system frequency when the system goes unbalanced [8].

FCDM system can be developed using smart meters and smart appliances. This kind of system has been reported in [9]. In the proposed system, some of the selected domestic loads are switched off using remote controlled plugs in order to provide primary frequency response. Domestic appliances are categorized into 5 groups, in the literature [9]. Depending upon the characteristic of loads they have individual affects on the primary demand response. The smart meters are used to measure the demand side frequency. Appliance groups are selected and switched off depending on the frequency drop. However in the proposed system, the appliances are completely switched off for different time schedules. Another study has described the method of dynamic demand control (DDC) [10]. DDC can be connected to refrigerators, freezers, air conditioner, and for some heating systems. DDC can switch on and off temperature settings of the device. Even though DDC system has its own operation techniques, they can be combined with the smart metering systems to increase the performances.

6.6 Enhance the HVAC System Performance

HVAC system stands for heating, ventilation, and air conditioning. HVAC systems are typically used at large buildings because of the high efficiency than the individual air-conditioning units and the cost effectiveness. These systems maintain a high quality in indoor air and control the temperature inside the building [11]. Recent studies show that the integration of smart meters into building management systems can be used to improve the system performance of HVAC systems. The smart meters can be used to predict the internal zone temperatures and measure the cooling capacity. Therefore with the aid of smart meters, HVAC loads can be managed in an effective way to cut down the peaks in the demand. There should be a non-intrusive load monitoring system, load operating schedules, and several environmental sensors to monitor and control the HVAC system [12].

6.7 Worldwide Load Control Programs

A pilot program was conducted with residential customers by ETSA, an electricity distributor in the State of South Australia. The objective was to reduce the summer peak by residential air conditioner cycling. Small direct load controllers were attached to the external compressors of the air conditioners. Signals were sent to these devices through a public radio station. They identified that effective load reduction requires a random overlapping switching program [6].

In the United States, Long Island Power Authority (LIPA) has developed a program to control small commercial air-conditioning thermostats. The objective was to reduce the peak load via load control. A device called "Programmable Comfort Choice thermostat" was used to control the air-conditioning units. The signals were sent to the pager tower via satellite, and then transmitted to the thermostats. Thermostats were able to collect the temperature in every minute. Power consumption was measured hourly and the data were reported back to the utility. LIPA expected 24.9 MW of peak demand reduction from 23,400 controlled air-conditioners [6].

A demand response spinning reserve pilot project is reported in [13]. This was tested in a feeder in California, USA, using the air conditioning load cycling program. Load shedding was triggered by sending a signal from the control center.

The California Public Utilities Commission approved a pilot program called automated demand response system (ADRS). A system with two-way communication was installed at consumer homes to control the home appliances. This was a web based home climate control system. The home owners could set climate control and pool pump runtime preferences both locally and remotely. Meanwhile the owners were able to view the current electricity price via internet or at the thermostat. Substantial peak load reduction was reported in this pilot project [6].

6.8 Optimal Energy Management

Load management and demand control systems are designed to reduce the peak demand, stabilize the system frequency, and save the electricity cost. When it comes to electricity savings the consumers can change the consumption patterns under different tariff schemes [14]. Modern smart metering systems support following tariff schemes

- Real time pricing (RTP): Price signals are generated in hourly basis or half hourly basis, in most cases reflecting the true cost of electricity. Consumers are updated in regular intervals (hourly of half hourly) by sending price signals to the smart meters or home display units.
- Time of use pricing (TOU): Designed considering the time of the day or the seasonal variations of electricity demand. Typically the 24 h are divided into several groups considering the peak and off peak demands. This tariff is fixed and pre defined compared to RTP, but can be changed yearly or half yearly.
- Critical peak pricing: The tariff is same as the TOU but higher rates are applied at higher demands. Prior to applying the higher rates, the consumers are notified.
- Day ahead pricing: This is similar to RTP, but the price schedule is sent to the consumer 24 h before the rates are applied. Generally, different prices are defined for different time locations (on hourly basis most of the time) of the next day.

Several studies reveal that the real time pricing has many advantages compared to other tariff schemes [14, 16]. The advantages might be economical, environmental, and energy efficient improvements [14]. RTP can provide many opportunities to reduce the electricity expenditure for the consumer. However there are several barriers to achieve the maximum benefits through RTP. Lack of home automation systems and lack of consumer awareness are the main barriers which have been identified according to Ref. [14]. Nevertheless these obstacles can be eliminated by introducing smart meters, smart appliances, and demand management systems. Modern smart metering systems support demand management and energy management which can be used for residential energy optimizations.

A demand response smart controller has been developed in [15], which can be easily implemented in residential energy management systems. This device can be mounted outside the electric appliance. Price signals are imposed from the smart meter and electricity consumption is limited when the price is high. With these functionalities, this device can run the residential lighting and other power devices to minimize the cost [15].

A mathematical model has been implemented for optimal and automated residential energy consumption scheduling in Ref. [16]. Here the minimization of electricity cost and the minimization of waiting time are discussed. The proposed system can be developed using smart meters with real time tariff combined with inclining block rates. On the other hand this system will help the power utilities to improve the load factor of the system [16].

An optimization model has been proposed using linear programming algorithm for load management in Ref. [17]. The electricity consumption is maximized subject to a minimum daily consumption level, maximum and minimum hourly load levels, and ramping limits. This system provides opportunities for residential or small business energy management. However the proposed system can only be used to adjust the hourly load level under hourly pricing scheme [17].

New approaches for optimal energy management through plug-in hybrid vehicles are still being introduced. An optimization framework for charging electricity vehicles is described in Ref. [18]. Another optimization methodology for charging and discharging of plug-in hybrid electric vehicles (PHEVs) is introduced in Ref. [14]. The retailer energy cost is minimized by controlling the charge and discharge of PHEVs. The customer behavior under time variant tariff schemes is also discussed with economic demand models [14].

References

1. Bruno S, Lamonaca S, Scala ML, Rotondo G, Stecchi U (2009) Load control through smart-metering on distributed networks. In: Proceedings IEEE powertech conference, Bucharest, June 28–July 2 2009
2. Distributed generation (2013) http://www.en.wikipedia.org/wiki/Distributed_generation. Accessed 18 Mar 2013
3. Lee PK, Lai LL (2009) A practical approach for smart meter applied in demand side and distributed generation. In: Proceedings advances in power system control, operation and management international conference, Hong Kong. 8–11 Nov 2009
4. Net metering (2012) http://www.en.wikipedia.org/wiki/Net_metering. Accessed 21 July 2012
5. Gao C, Redfern MA (2011) A review of voltage control in smart grid and smart metering technologies on distribution networks. In: Proceedings 46th international universities' power engineering conference, Soest, 5–8 Sept 2011
6. Crossley D (2008) The role of advanced metering and load control in supporting electricity networks. www.ieadsm.org/DownloadFile.aspx?q.../bycdjGyWJYla/. Accessed 22 July 2012
7. Whitfield A, Kemp A, Young D, D'Souza T, Lowe K, Hedynach Y (2008) Cost benefit analysis of smart metering and direct load control. sedc-coalition.eu/.../uploads/.../NERA-08-02-29-Cost-Benefit-Analysis
8. Vijayananda WMT, Samarakoon K, Ekanayake J (2010) Development of a demonstration rig for providing primary frequency response through smart meters. In: Proceedings universities power engineering conference, Cardiff, Aug 31–Sept 3 2010
9. Samarakoon K, Ekanayake J (2009) Demand side primary frequency response support through smart meter control. In: Proceedings universities power engineering conference, Glasgow, 1–4 Sept 2009
10. Short J, Infield DG, Freris LL (2007) Stabilization of grid frequency through dynamic demand control. IEEE Trans Power Systems J 22:1284–1293. doi:10.1109/TPWRS.2007.901489
11. HVAC control system (2013) http://www.en.wikipedia.org/wiki/HVAC_control_system. Accessed 18 March 2013
12. Sinha MN, Cox RW (2011) Improving HVAC system performance using smart meters. In: Proceedings energytech conference, Cleveland, 25–26 May 2011
13. Demand response spinning reserve demonstration (2012) http://www.certs.lbl.gov/pdf/62761.pdf Accessed 22 July 2012
14. Doostizadeh M, Khanabadi M, Esmaeilian A, Mohseninezhad M (2011) Optimal energy management of a retailer with smart metering and plug-in hybrid electrical vehicle. In: Proceedings environment and electrical engineering international conference, Rome, 8–11 May 2011
15. In-Ho Choi J, Lee H (2011) Development of a smart controller with demand response for AMI connection. In: Proceedings control automation and systems international conference, Gyeonggi-do, 27–30 Oct 2010
16. Mohsenian- Rad AH, Garcia AL (2010) Optimal residential load control with price prediction in real-time electricity pricing environments. IEEE Trans Smart Grid 1:120–133. doi: 10.1109/TSG.2010.2055903

17. Conejo AJ, Morales JM, Baringo L (2010) Real-time demand response model. IEEE Trans Smart Grid 1:236–242. doi:10.1109/TSG.2010.2078843
18. Kristoffersen TK, Capion K, Meibom P (2011) Optimal charging of electric drive vehicles in a market environment. J Appl Ener 88(5):1940–1948

Appendix 1
Schematic of Energy Measurement Chip

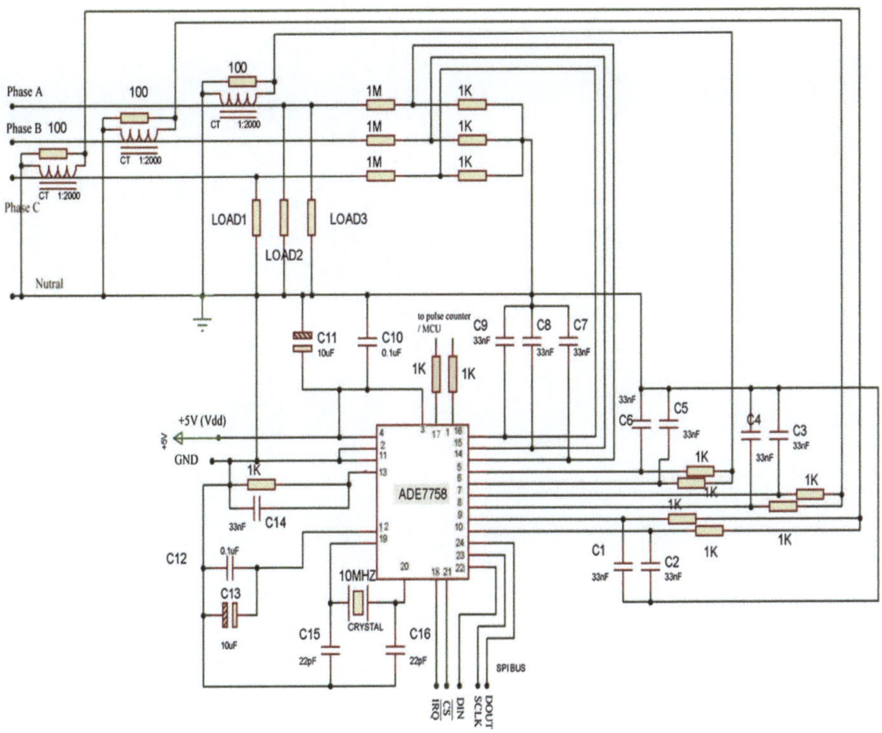

K. S. K. Weranga et al., *Smart Metering Design and Applications*,
SpringerBriefs in Applied Sciences and Technology,
DOI: 10.1007/978-981-4451-82-6,

Appendix 2
Schematic of PIC 18F452/ RTC/ LCD

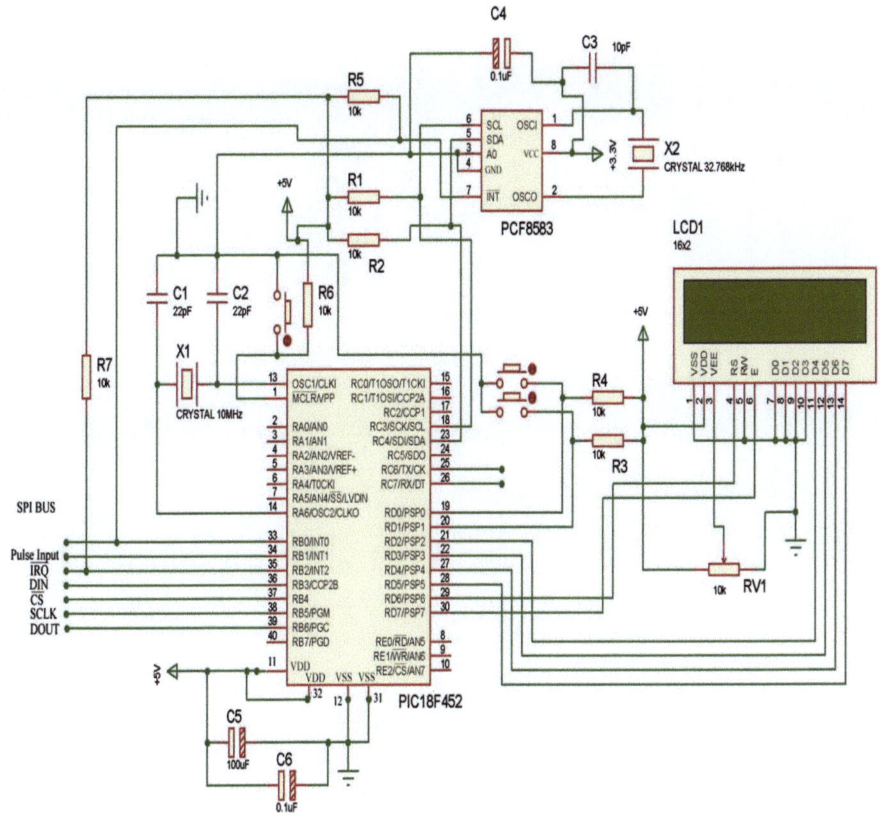

K. S. K. Weranga et al., *Smart Metering Design and Applications*,
SpringerBriefs in Applied Sciences and Technology,
DOI: 10.1007/978-981-4451-82-6, © The Author(s) 2014

Appendix 3
Schematic of SIM900 GSM Module

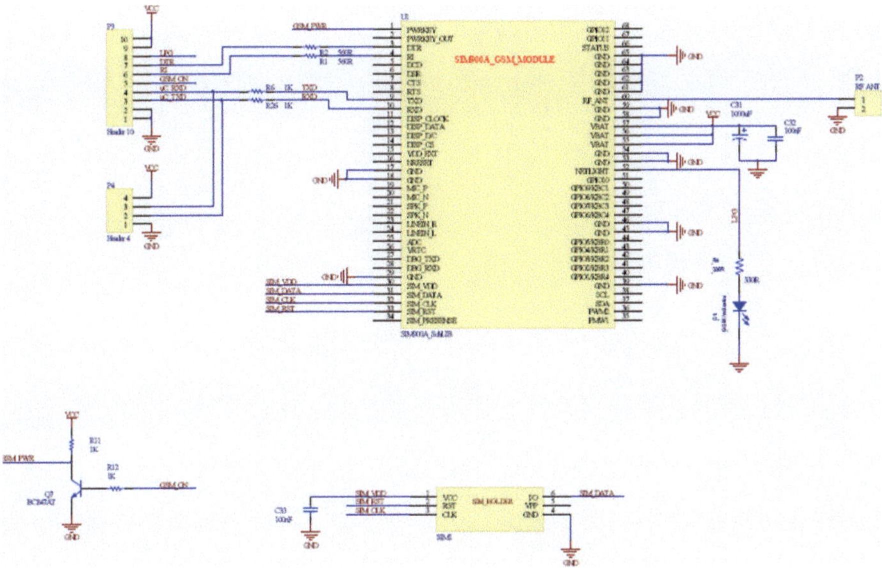

K. S. K. Weranga et al., *Smart Metering Design and Applications*,
SpringerBriefs in Applied Sciences and Technology,
DOI: 10.1007/978-981-4451-82-6, © The Author(s) 2014

Appendix 4
Schematic of Power Supply

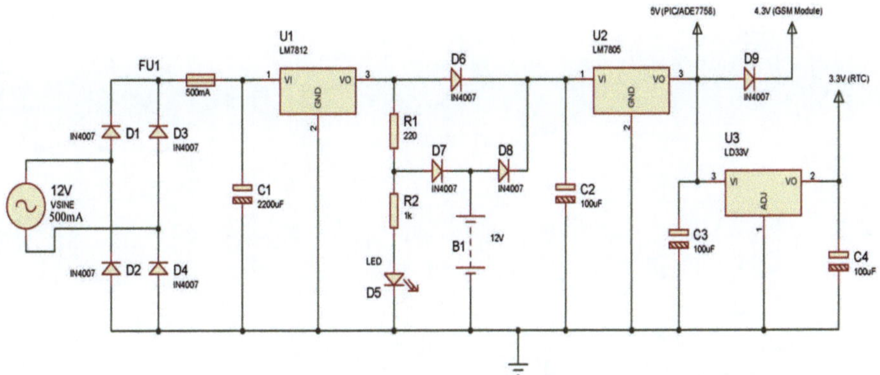

K. S. K. Weranga et al., *Smart Metering Design and Applications*,
SpringerBriefs in Applied Sciences and Technology,
DOI: 10.1007/978-981-4451-82-6, © The Author(s) 2014

Appendix 5
Some Useful Functions for Software Development

Note: Even though these functions are written for PIC microcontrollers using mikroC PRO 3.2v langue, they can be modified for other micro controller programming languages.

K. S. K. Weranga et al., *Smart Metering Design and Applications*,
SpringerBriefs in Applied Sciences and Technology,
DOI: 10.1007/978-981-4451-82-6, © The Author(s) 2014

//SPI write function//

```
int i; // Global variable

void writeade (unsigned short address, unsigned short size, unsigned short data2,unsigned short data1, unsigned
short data0)
{                                    // address, size of register, most byte, medium byte, least byte
address.F7=1;                        //MSB=1, write operation initialization for communication
register
PORTB.F3=0;                          //Chip select (CS) put to logic zero
Delay_us(1);

for(i=7;i>=0;i--)                    // write the specific address
{
 PORTB.F4=1;                         // clock pulse rising edge
 PORTB.F2=address.f7;                //write 1 to indicate that the next data transfer is write
 Delay_us(1);
 PORTB.F4=0;                         // clock pulse falling edge
 Delay_us(1);
 address=address<<1;                 // write the address by shifting next 7bits on communications register
}

        if(size>16)                  // if 3bytes to be written ( 24 bit register)
        {
                for(i=7;i>=0;i--)
                {
                PORTB.F4=1;          //clock pulse rising edge
                PORTB.F2=data2.f7;   // write the MSB of data2 in to DIN
                Delay_us(1);         // clock time
                PORTB.F4=0;          //clock pulse falling edge
                Delay_us(1);         // clock time
                data2=data2<<1;      // left shift operation on data2
                }
        }

        if(size>8)                   // if 2bytes to be written ( 16 bit register)
        {
                for(i=7;i>=0;i--)
                {
                PORTB.F4=1;
                PORTB.F2=data1.f7;
                Delay_us(1);
                PORTB.F4=0;
                Delay_us(1);
                data1=data1<<1;
                }
        }
```

```
        for(i=7;i>=0;i--)                     // if 1 bytes to be written ( 8 bit register)
        {
        PORTB.F4=1;
        PORTB.F2=data0.f7;
        Delay_us(1);
        PORTB.F4=0;
        Delay_us(1);
        data0=data0<<1;
        }

PORTB.F2=1;                  // Chip select (CS) put to logic one to finish the write operation

}
```

//SPI write function call example//

```
writeade(0x18,24,0,0b11000000,0b00111000);  // Interrupt Mask (0x18 of address) 24 bit length register (most
                                                byte=0, medium byte= 0b11000000, least byte =0b00111000)

Delay_us(2); // set small delay to stabilize the serial port

writeade(0x16,8,0,0,0b00011100);   // disable Phase B and Phase C contribution to APCF and VARCf
Delay_us(2);

writeade(0x13,8,0,0,0b00000000);   // enable APCF and VARCF by writing 0 to 0x13 register
Delay_us(2);
```

```
///////////////////////////////////////////////////////SPI read function///////////////////////////////////////////////////////

unsigned short r2data=0; r1data=0; r0data=0; // Global variables r2data is first 8 bits, r1data is next 8 bits, and r0data
                                                          is last 8 bits of a 24 bit register
int i; // Global variables

void readade (unsigned short address, unsigned short size)
{                                                              //address, size of register

address.F7=0;                          //MSB=0, read operation initialization for communication
PORTB.F3=0;                            //Chip select (CS) put to logic zero

Delay_us(1);

for(i=7;i>=0;i--)                            // write the specific address
{
PORTB.F4=1;                                  // clock pulse rising edge
PORTB.F2=address.f7;                 //write 0 to indicate that the next data transfer operation is read
Delay_us(1);
PORTB.F4=0;                          //clock pulse falling edge
Delay_us(1);
address=address<<1;                  // write the address by shifting next 7bits on communications register
}

Delay_us(5);

        if(size>16)             // read first 8 bits
        {
                for(i=7;i>=0;i--)
                {
                r2data=r2data<<1;       shift next 7bits on r2data
                PORTB.F4=1;
                Delay_us(1);
                r2data.f0=PORTB.F5;  // read a bit from DOUT
                PORTB.F4=0;
                Delay_us(1);
                }
        }


        if(size>8)              // read next 8 bits
        {
                for(i=7;i>=0;i--)
                {
                r1data=r1data<<1;
                PORTB.F4=1;
                Delay_us(1);
                r1data.f0=PORTB.F5;
                PORTB.F4=0;
                Delay_us(1);
                }
        }
```

```
for(i=7;i>=0;i--) // read last 8 bits
{
r0data=r0data<<1;
PORTB.F4=1;
Delay_us(1);
r0data.f0=PORTB.F5;
PORTB.F4=0;
Delay_us(1);
}

PORTB.F3=1; // Chip select (CS) put to logic one to finish the read operation
}
```

//SPI read function call example//

```
readade(0x0D,24); // read 24 bit AVRMS register of address 0x0D

readade(0x01,16); // read 16 bit AWTTHR register of address 0x01

readade(0x10,12); // read 12 bit FREQ register of address 0x10
```

///Binary to Decimal conversion function///

```
float decimal, r2dataa, r1dataa, r0dataa;

float B2D()
{
        if(r2data.F7==1)                                   // MSB of first byte is one, negative value
        {
        r2dataa=~r2data;
        r1dataa=~r1data;
        r0dataa=~r0data;
        decimal=((r0dataa)+(r1dataa)*256+(r2dataa)*256*256+1)*(-1);      // decimal negative value
        }

        else                                               // positive value
        {
        r2dataa=r2data;
        r1dataa=r1data;
        r0dataa=r0data;
        decimal=r0dataa+(r1dataa*256)+(r2dataa*65536);                   // decimal positive value
        }

return decimal;

UART1_Write_Text("ATZ\r\n");
```

```
/////////////////////////////////////////////SIM 900 AT commands for SMS/////////////////////////////////////////////
UART1_Init(19200);          // initialize UART module at 19200 bps
Delay_ms(100); // small time delay to stabilize the serial port
UART1_Write_Text("AT+CMGF=1\r\n"); // SMS write mode
Delay_ms(100);
UART1_Write_Text("AT+CMGS=\"0123456789 \"\r\n"); // specify the phone number example 0123456789
Delay_ms(100);
UART1_Write_Text("abcdef"); // write any text example abcdef
Delay_ms(100);
UART1_Write(0x0D); // hex for "enter"
Delay_ms(100);
UART1_Write(0x1A); // send the SMS
Delay_ms(50000);
```

Appendix 6
Mathlab Code for Simulation of Fig. 5.5

Note: Software is shown only for first 15 minutes to demonstrate the algorithm.

K. S. K. Weranga et al., *Smart Metering Design and Applications*,
SpringerBriefs in Applied Sciences and Technology,
DOI: 10.1007/978-981-4451-82-6,

```
%%%%%%%%%%%%%%%%%%%%%%%%%%%%%%%%Simulation%%%%%%%%%%%%%%%%%%%%%%%%%%%%%%%%%%%

Du= 1000; % User defined demand level
t=1:1:15; % time from t=1 to t=15 minutes
T1=0:0.1:10; % discrete time used for A0 area calculation
Ta1=0:0.1:15; % discrete time used for A1 area calculation

A=1000*[0.868945,   0.897963,   0.925331,   0.951105,  0.975343,   0.998101,
1.019434,   1.039395, 1.058039,   1.075416,   1.091579,   1.106578,
1.120462,   1.133280,   1.145079]; % take first 15 sample points

for i= 1:15 % plot demand against time for 15 minutes

    plot(t(1,i),A(i), 'o','MarkerFaceColor','r','MarkerSize',3);
    xlabel('Time (minutes) ','LineWidth' ,4);
    ylabel('Demand (kVA) ','LineWidth' ,4);
    title('Demand forecast Simulation');
    pause(0.5);
    hold on;

if(i==10) % at the 10th sample point forecast the demand in next 5 minutes

    a4=0.0026*(A(1))-0.0032*(A(2))- 0.0024*(A(3))+0.0004*(A(4))+
0.0026*(A(5))+ 0.0026*(A(6))+0.0004*(A(7))-  0.0024*(A(8))-0.0032*(A(9))+
0.0026*(A(10)); % calculate the coefficients
    a3=-0.06585*(A(1))+0.07323*(A(2))+0.06128*(A(3))- 0.00359*(A(4))-
0.05536*(A(5))-0.06002*(A(6))-0.01563*(A(7))+0.04768*(A(8))+0.06779*(A(9))-
0.04953*(A(10)); % calculate the coefficients
    a2=0.5795*(A(1))-0.5533*(A(2))-0.5147*(A(3))-
0.0403*(A(4))+0.3685*(A(5))+ 0.4455*(A(6))+ 0.1583*(A(7))- 0.2903*(A(8))-
0.4635*(A(9))+ 0.3103*(A(10)); % calculate the coefficients
    a1=-2.0792*(A(1))+1.4914*(A(2))+1.6184*(A(3))+ 0.3727*(A(4))-
0.8158*(A(5))-1.1585*(A(6))- 0.5073*(A(7))+0.6443*(A(8))+1.1623*(A(9))-
0.7284*(A(10)); % calculate the coefficients
    a0=2.4999*(A(1))-0.8333*(A(2))- 1.2499*(A(3))- 0.4166*(A(4))+
0.4999*(A(5))+0.8333*(A(6))+0.4166*(A(7))-0.4166*(A(8))-0.8333*(A(9))+
0.4999*(A(10)); % calculate the coefficients
    f1= a4*(t1.^4) + a3*(t1.^3)+ a2*(t1.^2)+ a1*t1 + a0; %derive the demand
function in first 15 minutes
    A1=18.38213869*(A(1))- 27.90768259*(A(2))-
16.7546863*(A(3))+10.406347059*(A(4))+26.872328887*(A(5))+
20.67186248*(A(6))-5.434756927*(A(7))-33.955542226*(A(8))-
32.666814373*(A(9))+45.386797580*(A(10)); % Calculate the forecasted area for
next 5 minutes
    Y1= -0.000039*(T1.^4) +0.010026*(T1.^3)- 0.884614*(T1.^2)+ 31.602446*T1 +
838.2172; % polynomial demand variation is assumed for first 10 minutes
```

```
/This is only a demonstration and actual area is calculated in smart meters
using a higher sampling rate

    A0=trapz(T1,Y1);% calculate the area of first 10 minutes

    MDE1=(A0+A1)/15 ;% Calculate the forecasted average demand at 15 minutes

  end


if((i==1)||(i==2)||(i==3)||(i==4)||(i==5)||(i==6)||(i==7)||(i==8)||(i==9)||(i
==10))
    plot(t(1,i),A(i), 'O', 'MarkerFaceColor','g','MarkerSize',3); % show
sample points
    pause(0.5);
 end

 hold on;

 if((i==11)||(i==12)||(i==13)||(i==14)||(i==15))
 plot(t(1,i),f1(1,i), 'O', 'MarkerFaceColor','b','MarkerSize',3); % show
forecasted demand points
 pause(0.5);
 end

 hold on;

 if(i==10)
     if(MDE1>Du) % compare the forecasted average demand and user defined at
t=10 minutes demand
     text(t(1,i),A(i),'\leftarrow warning',...
   'HorizontalAlignment','left');
     end                             % show warning message on figure

 end
 hold on;

drawnow
end
```

MIX
Papier aus verantwortungsvollen Quellen
Paper from responsible sources
FSC® C105338

If you have any concerns about our products,
you can contact us on
ProductSafety@springernature.com

In case Publisher is established outside the EU,
the EU authorized representative is:
Springer Nature Customer Service Center GmbH
Europaplatz 3, 69115 Heidelberg, Germany

Printed by Libri Plureos GmbH
in Hamburg, Germany